Topics in Recreational Mathematics 2/2016

Editor-in-chief

Charles Ashbacher
5530 Kacena Ave
Marion, IA 52302 USA

cashbacher@yahoo.com

Artwork

Caytie Ribble

Problems

Lamarr Widmer

Contributor

Rachel Pollari

ISBN-13: 978-1534964846

CONTENTS

Note From the Editor

 Welcome to the latest installment of my **Topics in Recreational Mathematics** series. This is the second book of 2016 and the seventh in the overall series. It is behind schedule due to the death of my uncle and having to deal with some major issues with family property.

 A new feature in this installment is a short section on wordplay, most of which was written by contributor Rachel Pollari. In it, famous sayings are overworded in a humorous way. While it is not strictly speaking within the usual classification of recreational mathematics, I find it a great deal of fun. Hope you enjoy it.

 There are solid plans to publish two additional books in this series in 2016 and I have in fact started working on them as well. Some of the regular features naturally extend across the issue borders. I am also working on the development of other book projects involving content that appeared in **Journal of Recreational Mathematics**.

 As always, I welcome feedback of all kinds.

Charles Ashbacher

cashbacher@yahoo.com

CAYTiEDiD SeQuence

GREEK SHEEP

Wordplay

Rachel Pollari

Charles Ashbacher

The following were all written by Rachel Pollari.

Bread ends masticated in amity are more restorative than a banquet occupied in apprehension.

Avians divvying the same plumage congregate collectively.

Moving leisurely but deliberately will triumph in contest.

Take observation before you hurdle exuberantly.

Vehement persuasion is not a tonic.

In squabbling about the sun's dark silhouette, we regularly misplace the matter.

It is more apt to be supple than to snap.

Formerly chomped, doubly wary.

One bloke's animal muscle is another chap's Kryptonite.

One fluttery avian does not signify the post-spring season.

Afford the branch, cosset the offspring.

Those who pursue universal satisfaction satisfy none.

What is conceived within marrow will fuse to the epidermis.

Malevolent aspirations, like fowl, return to one's dwelling for shelter.

Obligation is the dam of innovation.

A shady enticement grasped denotes a rogue in the pulse.

Any justification will oblige a despot.

Beware lest you mislay the constituent by snatching at the shade.

Do not compute your poultry afore they emerge.

Don't bawl over splattered udder emissions.

It is not merely sublime plumage that marks superlative fowls.

The last one was written by Charles Ashbacher

An avian contained by phalanges economically dominates two fowl found in diminutive topiary

Is There Safety in Numbers? A Study of California's Sanctuary Cities

Paul M. Sommers
Economics Department
Middlebury College
Middlebury, Vermont 05753
psommers@middlebury.edu

Abstract

The author examines eight different crime rates (three defined for crimes against persons, five for crimes against property) in sanctuary cities to the corresponding average of all other cities in California of comparable size. The results indicate that crime rates in 2013 were significantly higher in all sanctuary cities and, in particular, for California's fifteen cities with a population over 75,000, four crime rates — robbery, motor vehicle theft, aggravated assault, and murder — were significantly higher in sanctuary cities.

If you're going to San Francisco
Be sure to wear some flowers in your hair.
If you're going to San Francisco
You're gonna meet some gentle people there.

— "San Francisco" (1967)
Written by John E. A. Phillips
Sung by Scott MacKenzie

There's no safety in numbers,
or in anything else.

— James Thurber [1]
The moral to the fable,
"The Fairly Intelligent Fly"

Moving to California? Immigrants may take up residence in cities which have adopted "sanctuary" policies, so-called because city employees are not obligated to notify the federal government of the presence of undocumented immigrants living in their community. The vitriolic rhetoric among 2016 presidential candidates over illegal immigration notwithstanding, one might wonder what *are* the statistical differences (if any) between crime rates in sanctuary cities and all other cities.

In this brief research note, we examine eight different crimes — murder, rape, robbery, aggravated assault, burglary, larceny-theft, motor vehicle theft, and arson — in California's sanctuary cities and all other cities in the Golden State of comparable size.

Table 1 lists thirty-two sanctuary cities in California.[1] All of California's 462 cities are divided into four groups, by population. The dividing points closely correspond to quartiles.[2] The FBI's Uniform Crime Reports [3] lists the population and the number of offenses for crimes against persons (murder and non-negligent manslaughter, rape, and aggravated assault) and property (robbery, burglary, larceny-theft, motor vehicle theft, and arson) for each of California's 462 cities. Crimes as of 2013 are expressed as rates per 100,000 inhabitants.

For each of the four population groups, a series of two-sample *t*-tests were run comparing the average crime rate in sanctuary cities to the corresponding average of all other cities in California of comparable size. The results are presented in Table 2.

For the smallest group (cities with a population under 12,500), most crime rates (excluding murder) are, on average, significantly higher in the three sanctuary cities of Industry, Sonoma, and Vernon. For the next two groups of cities, that is, California's cities ranging in population from 12,500 to 75,000, robbery and motor vehicle theft crime rates are, on average, significantly higher in sanctuary cities. And, for the largest group (cities with populations over 75,000), not only are robbery and motor vehicle theft crime rates, on average, significantly higher in sanctuary cities, but so too are several rates involving crimes against persons, namely, aggravated assault and murder.

Concluding Remarks

A comparison of crime rates in California's sanctuary cities and all other cities of comparable size shows that in 2013 robbery and motor vehicle theft were, on average, significantly higher in sanctuary cities. And, as for the most populous cities in the state (that include the sanctuary cities of Los Angeles, San Diego, San Jose, and San Francisco), four crime rates were, on average, significantly higher in sanctuary cities: robbery, motor vehicle theft, aggravated assault, and murder.

Interested readers might be curious to run similar tests for U.S. cities in other states bordering Mexico (Texas, Arizona, and New Mexico) or in any of the sixteen U.S. states bordering Canada.[3]

Table 1
California's Sanctuary Cities,
by Population, 2013

Population under 12500
> City of Industry (223)
> Sonoma (10912)
> Vernon (115)

Population between 12500 and 32000
> Commerce (12995)
> Greenfield (16922)
> Maywood (27744)

Population between 32001 and 75000
> Bell Gardens (42980)
> Coachella (43330)
> Cypress (49067)
> Davis (66126)
> Lynwood (71077)
> Montebello (63566)
> National City (59637)
> Paramount (54868)
> Pico Rivera (63710)
> Santa Cruz (62517)
> Watsonville (52076)

Population 75000 or over
> Downey (113222)
> Fresno (508876)
> Lakewood (81086)
> Long Beach (469665)
> Los Angeles (3878725)
> Norwalk (106518)
> Oakland (403887)
> Richmond (107341)
> San Bernardino (214322)
> San Diego (1349306)
> San Francisco (833863)
> San Jose (992143)
> Santa Clara (120150)
> Santa Maria (102051)
> South Gate (95591)

Source: California's sanctuary cities listed at http://www.ojjpac.org/sanctuary.asp .

Table 2
Crime Rates in California Cities, 2013,
by Sanctuary Status and Population[a]

Population	Average rates (per 100,000)		*p*-value on difference[b]
	Sanctuary cities	All other cities	
Cities under 12500 ($n_1 = 3$, $n_2 = 115$)[c]			
Murder	0	1.81	.585
Rape	607.07	30.26	**<.001**
Robbery	6492.27	52.77	**<.001**
Aggravated assault	7016.83	278.86	**<.001**
Burglary	24750.83	751.29	**<.001**
Larceny-theft	160008.70	1960.38	**<.001**
Motor vehicle theft	64537.00	283.52	**<.001**
Arson	152.53	21.58	**<.001**
Cities between 12500 and 32000 ($n_1 = 3$, $n_2 = 110$)			
Murder	9.34	3.68	.136
Rape	11.91	20.82	.378
Robbery	199.53	78.56	**.002**
Aggravated assault	301.97	236.89	.586
Burglary	443.67	730.76	.286
Larceny-theft	1757.33	1515.70	.662
Motor vehicle theft	984.33	289.12	**<.001**
Arson	23.70	20.19	.821
Cities between 32001 and 75000 ($n_1 = 11$, $n_2 = 110$)			
Murder	3.77	2.14	.101
Rape	21.71	15.44	.107
Robbery	140.36	81.04	**.002**
Aggravated assault	204.39	158.69	.818
Burglary	618.39	566.09	.551
Larceny-theft	1575.82	1565.57	.965
Motor vehicle theft	583.09	301.81	**<.001**
Arson	20.64	13.17	.162

Table 2
Crime Rates in California Cities, 2013,
by Sanctuary Status and Population[a]
(Continued)

| Population | Average rates (per 100,000) | | p-value on difference[b] |
	Sanctuary cities	All other cities	
Cities over 75000 ($n_1 = 15$, $n_2 = 95$)			
Murder	7.44	3.80	**.017**
Rape	22.18	18.33	.181
Robbery	275.69	119.12	**<.001**
Aggravated assault	294.90	199.28	**.024**
Burglary	721.21	585.61	.150
Larceny-theft	1906.07	1657.63	.143
Motor vehicle theft	773.87	397.33	**<.001**
Arson	20.44	15.21	.221
All cities ($n_1 = 32$, $n_2 = 430$)			
Murder	5.66	2.81	**.004**
Rape	75.89	21.42	**<.001**
Robbery	804.84	81.26	**<.001**
Aggravated assault	894.63	219.80	**<.001**
Burglary	2912.62	662.06	**<.001**
Larceny-theft	16600.72	1678.74	**<.001**
Motor vehicle theft	6705.81	314.77	**<.001**
Arson	33.20	17.67	**.011**

[a]Crime rates per 100,000 population are computed using population figures and the number of offenses at https://www.fbi.gov/about-us/cjis/ucr/crime-in-the-u.s/2013/crime-in-the-u.s.-2013/tables/table-8/table-8-state-cuts/table_8_offenses_known_to_law_enforcement_california_by_city_2013.xls.
[b]The p-value reported is for a two-tailed t-test.
[c]The number of sanctuary cities is denoted by n_1 and the number of all other California cities of comparable size is denoted by n_2.

References

1. J. Thurber, *Fables for Our Time*. Harper & Row, Publishers, 1939. Perennial Library edition first published in 1974.

2. S. Salvi. "Sanctuary Cities: What Are They?" 2015. Retrieved from http://www.ojjpac.org/sanctuary.asp.

3. California's offenses known to law enforcement by city, 2013. Retrieved from https://www.fbi.gov/about-us/cjis/ucr/crime-in-the-u.s/2013/crime-in-the-u.s.-2013/tables/table-8/table-8-state-cuts/table_8_offenses_known_to_law_enforcement_california_by_city_2013.xls.

Footnotes

1. See S. Salvi [2] and the list of sanctuary cities for the state of California.

2. The quartiles are simply the medians of the two halves of the list of 462 cities, by population. The first quartile is 12168; the second quartile or median is 32046; and the third quartile is 72171. The min is 115 (Vernon, a sanctuary city); the max is 3878725 (Los Angeles, also a sanctuary city).

3. The sixteen states bordering Canada are Maine, New Hampshire, Vermont, New York, Pennsylvania, Ohio, Michigan, Illinois, Indiana, Wisconsin, Minnesota, North Dakota, Montana, Idaho, Washington, and Alaska.

Some Notes on the Evaluation of Magic Sums of Magic Squares

Hossein Behforooz
hbehforooz@utica.edu
www.utica.edu/hbehforooz

Abstract

We have a formula to calculate the magic sum of magic squares with positive consecutive integers. But we do not have a formula when entries of the magic squares are not consecutive positive integers. In this paper, we have some comments about this situation.

Background and Motivations

Recently, in a family dinner table conversation in a party when I was trying to amuse and entertain people with interesting properties of magic squares by showing them very simple three by three or four by four famous magic squares, my nephew Reza Zali who is a chemical engineer, made a comment. He mentioned that the magic sums in those simple examples were the average of two sums of the first and last n numbers of the magic squares. His comment became my main motivation to investigate this issue and was the motivation for this paper.

On the Evaluation of Magic Sums

We know that magic squares are square tables of numbers (usually with positive integers) such that the sum of numbers along the rows, columns and two diagonals are all equal to a number which is called magic sum or magic number or magic constant S. In almost every book about magic squares, we see formula (1) to help us to compute the magic sum of an n by n magic square with n^2 consecutive natural numbers $1, 2, 3, \dots, n^2$,

$$S = \frac{n(n^2+1)}{2}, \tag{1}$$

See for example the new book on magic square [5, page 205] "Before Sudoku, the World of Magic Squares" from two chemical engineers or [1, page 55] and [6 page 1].

For example the magic sum for 3 by 3 Lo Shu magic square is S = 15 = 3(9+1) / 2 and for a 4 by 4 magic square S = 34 = 4 (16 + 1) / 2 and for a 5 by 5, it is S = 65 = 5(25+1) / 2.

The proof of the above formula is not that difficult. The sum of all numbers from 1 to n^2 is $n^2(n^2 + 1)/2$ and when we divide this number by n to get the one n^{th} of this number for each row or column we obtain $S = n(n^2 + 1)/2$. In a similar manner, the magic sum for any magic cube with entries $1, 2, \dots, n^3$ is, (see for example [5, page38]),

$$S = \frac{n(n^3 + 1)}{2}.$$

The above formula $S = n(n^2 + 1)/2$ works only if the entries of the magic squares are consecutive natural numbers 1, 2, 3, ..., n^2. Otherwise this formula does not help to compute the magic sum of the magic square.

New Way to Compute the Magic Sums

Here comes a brand new magic way to compute the magic sum. Let's start with examples. From formula (1), we know that for any 3 by 3 magic square with entries 1, 2, 3, 4, 5, 6, 7, 8, 9, the magic number is $S = 3(9+1)/2 = 15$.

Now, we can get this number from a simple new technique,

$$S = [(1 + 2 + 3) + (7 + 8 + 9)] / 2 = 15.$$

This is the average of two sums of the first 3 smallest numbers 1, 2, 3, and the last 3 largest numbers 7, 8, 9. In any 4 by 4 magic square with entries 1, 2, 3, 4, ..., 15, 16, from the above formula we get $S = 4(16 + 1) / 2 = 34$. With new magic calculation, we obtain

$$S = [(1 + 2 + 3 + 4) + (13 + 14 + 15 + 16)] / 2 = 34.$$

Here is the main theorem or conjecture of this study.

Theorem 1: The magic sum S of any magic square of order n, is equal to the average of two sums S_1 and S_2, where S_1 is the sum of the first n smallest entries and S_2 is the sum of the last n largest entries of the magic square,

$$S = \frac{S_1 + S_2}{2}. \tag{2}$$

Proof: We prove this theorem for magic squares of order n with n^2 consecutive positive integers 1, 2, 3, ..., $(n^2 - 1)$, n^2. Here we have

$$S_1 = 1 + 2 + 3 + ... + (n - 1) + n = \frac{n(n+1)}{2} = \frac{n^2+n}{2},$$

$$S_2 = (n^2 - n + 1) + (n^2 - n + 2) + ... + (n^2 - 1) + n^2 =$$

17

$$= (n^2 - n + 1) + (n^2 - n + 2) + \cdots + (n^2 - n + n - 1) + (n^2 - n + n) =$$

$$= nn^2 - nn + \frac{n(n+1)}{2} = \frac{2n^3 - 2n^2 + n^2 + n}{2} = \frac{2n^3 - n^2 + n}{2}.$$

And from here, we have

$$\frac{S_1 + S_2}{2} = \frac{1}{2}\left[\frac{n^2 + n}{2} + \frac{2n^3 - n^2 + n}{2}\right] = \frac{2n^3 + 2n}{4} = \frac{n(n^2 + 1)}{2}.$$

So $S = \frac{S_1 + S_2}{2}$. I leave the proof for nonconsecutive numbers as an open problem to the interested readers. In the meantime, I have examined this claim for many magic squares and always the formula (2) worked very well with no problem. In other words, I couldn't find any counter example to reject or deny this theorem.

Demonstration: Let's consider the following famous magic square with prime numbers. Since magic squares are my favorite subject to study (see [4], the list of my published papers in magic squares) my colleagues in mathematic department at Utica College has presented this plaque to me when I left the department chair position in 2012.

The magic sum of this magic square is S = 807 and we can get it from (2)

$$S = \frac{S_1 + S_2}{2} = \frac{(71 + 149 + 191) + (347 + 389 + 467)}{2} = \frac{411 + 1203}{2} = 807.$$

As I mentioned earlier, our new formula (2) works for any magic square. We can use this formula to find the magic sum of the Behforooz Calendarical Magic Squares [2] by adding the dates of the first week and the dates of the last week. For more details on this, see [2].

A Curious Historical Note

 The magic square in the above plaque is part of a long story and I think it is worth mentioning this historical note here for the mathematical entertainment of a new generation of recreational mathematics readers. There is an interesting story behind the above magic square plaque that we can find in the literature, see for example, the web site [10] about Emperor Charlemagne (Reigned from December 25, 800 to January 28, 814).

Emperor Charlemagne had ordered a pentagonal fort to be built with five walls and five gates. As good luck charms to engrave on each wall, he wanted five third order magic squares all with prime numbers and with the same magic sum, see for example [7, page 273]. Somehow in those days without computer and mathematical software, mathematicians found the five magic squares with prime numbers and with one common magic sum S = 807 in figure 1. Interestingly, this number 807 is the middle number between two dates, the beginning date and terminal date of Emperor Charlemagne reign period, 807= (800 + 814)/2. Is this a coincidence? Is this part of the secrets of the magic squares?

More on Magic Sums

If n, the order of magic square, is odd and 1, 2, 3, ...,n^2 are the entries of the magic square, then $(n^2 + 1)/2$ is the middle number of the entries and its place in the table is always at the center cell of the table. So, in any magic square with odd order, the magic sum $S = n(n^2 + 1)/2$ is the product of n by the central entry. For example in the 3 by 3 Lo Shu magic square, 5 is at the central cell and S = 3 × 5 = 15. And in a 5 by 5 magic square with entries 1, 2, ..., 25, the central number is 13 and magic sum is S = 13 × 5 = 65. This idea is true even if the entries are not consecutive numbers. If we pay attention to the above five 3 by 3 magic squares, we notice that in each of those magic squares, the central number is the middle number between other entries and S = 3 × 269 = 807. I have checked this idea in many other magic squares with odd order and I couldn't find a counter example to reject this new idea. Here is another theorem without proof and I leave the proof to the interested readers again.

Figure 1

479	71	257
47	269	491
281	467	59

389	191	227
107	269	431
311	347	149

401	257	149
17	269	521
389	281	137

401	227	179
47	269	491
359	311	137

191	227	389
467	269	71
149	311	347

$$S = 807$$

Theorem 2: In any magic square with odd order $n = 2k + 1$, the central number is the middle number (median) of all entries and the magic sum is equal to the product of central number by n.

Proof: As I mentioned in [2], we are in the world of recreational mathematics and I prefer to avoid long and complicated proofs. Presenting a simple proof or a simple demonstration is enough. For this reason, we prove this theorem for 3 by 3 magic squares. Consider the following magic square with magic sum S.

a	b	c
d	e	f
m	n	p

We have

$a + e + p = S, \quad c + e + m = S, \quad d + e + f = S.$

By adding these three equations and rearranging the numbers we obtain

$(a + d + m) + (c + f + p) + 3e = 3S.$

Which implies that $3e = S$ and this shows that the magic sum S is 3 times the central number e. Since $d + e + f = S$ and $S = 3e$, we get $e = (d + f)/2$. That mean that the central entry e is the middle number and also the mean number for all two neighbor numbers of e, which are (d, f), (b, n), (a, p) and (c, m).

This proves that the two parts of theorem 2 are true for any 3 by 3 magic squares. For higher odd order magic squares, I have examined these claims and I found that the theorem 2 works very well. I couldn't find any counter example to reject the theorem.

Based on this theorem 2, those 45 prime numbers in the above mentioned five 3 by 3 magic squares are such that the central number 269 is the middle number for all other 44 prime numbers and those 44 prime numbers can be partitioned into five subsets each with eight primes and with the same sum 2152. After including 269 to each subset, we can arrange these primes and construct five 3 by 3 magic squares. The arrangement of the numbers to make the magic squares is the easy part of the job. It is similar to the construction of 3 by 3 magic squares by using the dates of the calendars, see [2]. That is why we must think twice to the well done job that those people did in years 800 to create those five magic squares. In May 1975, our pioneer recreational mathematician, Charles Wilderman Trigg, has proposed Problem MM 943 in May 1975 issue of MAA Mathematics Magazine [8] to reconsider and reconstruct the above five magic squares with the above conditions. The solution for this proposal from Bob Prielipp (also solved with some other readers) can be found in September 1976 issue of the same magazine [9]. Interestingly, the answer to Charlemagne's problem is the above five magic squares and there is no other solution. With including negative prime numbers, there is another set of five 3 by 3 magic squares with positive and negative prime numbers. But as we know, in those days the negative numbers were either unknown or immoral (see comments on this from Scott Smith in [9]). The bottom line is that the solving Charlemagne's problem in the year 800 is a remarkable result for their time with no computer, no mathematical software, no graphing calculator or no slide rule. Their mathematical tools were, at the most, a simple abacus to add or subtract the numbers. Even for me, 50 years ago, the abacus of my businessman father was the only tool to use to construct the Behforooz-Franklin 32 by 32 magic square and publish it later in [3] when I came to America. It was in the good old days when I was an undergraduate mathematics student in Tehran University, Iran and I was working on magic squares. Here is the picture of that abacus which is in base 10 (very different and simpler than those Chinese or oriental type abacuses in base 5) and in those days, these types of abacus were a very popular tool in Iran to compute and arrange the bills and statement in businesses, banks and taxation offices and also determine grade report cards in schools and universities.

Figure 2

Abacus in Base 10 Used in Iran Before Mechanical or Digital Calculators

References

1. W. H. Andrews, *Magic Squares and Cubes*, Dover, New York, 1917.

2. Hossein Behforooz, Behforooz Calendarical Magic squares, *Topics in Recreational Mathematics*, No. 3, 69-79, 2015.

3. Hossein Behforooz, Behforooz-Franklin 32 by 32 Magic Square, *Journal of Recreational Mathematics, Vol. 33, No. 2*, 107-110, 2004-2005.

4. Hossein Behforooz, Personal website at, www.utica.edu/hbehforooz

5. Seymour S. Block and Santiago A. Tavares, *Before Sudoku, the World of Magic Squares,* Oxford University Press, UK, 2009.

6. Clifford Pickover, *The Zen of Magic Squares, Circles, and Stars,* Princeton University Press, Princeton, NJ, 2002.

7. Stanley Rabinowitz and Mark Bowron, *Index to Mathematical Problems,* Vol. 2, 1975-1979, MathPro Press, MA, USA 1999.

8. Charles W. Trigg, Problem # MM 943, MAA Mathematics Magazine, Vol. 48, No. 3, page 181, May 1975.

9. Bob Prielipp, Solutions to Problem # MM 943, MAA Mathematics Magazine, Vol. 49, No. 4, page 213, September 1976.

10. https://en.wikipedia.org/wiki/Charlemagne

Figure 3

Behforooz-Franklin 32 by 32 Magic Square with Magic Sum S= 16400

Making the Grid Square-Free

Dr. Moloy De

demoloy@yahoo.co.in

Abstract

Considered is a finite regular grid of squares. The main question addressed is on the minimum number of edges needed to be removed to make the grid square-free.

Keywords: Lattice Grid, Matchstick Puzzle.

Introduction

There are $2n(n+1)$ edges in an $n \times n$ square grid containing $n(n+1)(2n+1)/6$ squares in total. Let $P(n)$ denote the minimum number of matchsticks needed to be removed to make the square grid free of squares.

Known values of $P(n)$ are summarized in table 1.

Table 1

n	$P(n)$
1	1
2	3
3	6
4	9
5	14

On the 3 × 3 Grid

A crude search algorithm is developed in R to find that it needs a minimum of 6 edges to be removed to make the 3×3 grid square-free. The algorithm runs out of memory when applied on 4×4 grid. The edges removed appear in figure 1.

Figure 1

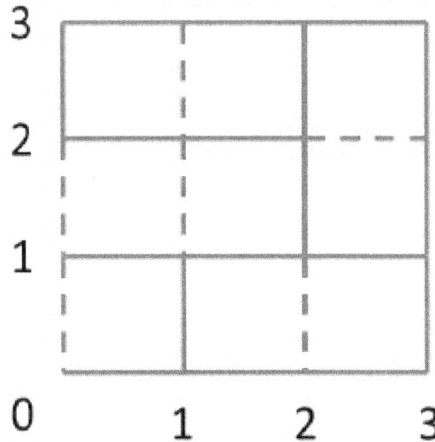

Finding Non-trivial Upper Bounds

Algorithm I

An iterative algorithm is developed in R where in each iteration the edge, removal of which removes the maximum number of squares, is removed. The observed upper bounds are summarized in table 2.

Table 2

n	Upper Bound of $P(n)$
1	1
2	3
3	6
4	10
5	15
6	22
7	28
8	39
9	49
10	57

Algorithm II

A second iterative algorithm is developed in R where bottom left corner squares are made square free successively. The observed upper bounds are summarized in table 3.

Table 3

n	Upper Bound of $P(n)$
1	1
2	3
3	6
4	10
5	15
6	21
7	28
8	36
9	45
10	55

Figure 2 shows the 10 × 10 grid with the edges removed that will make it square-free.

Figure 2

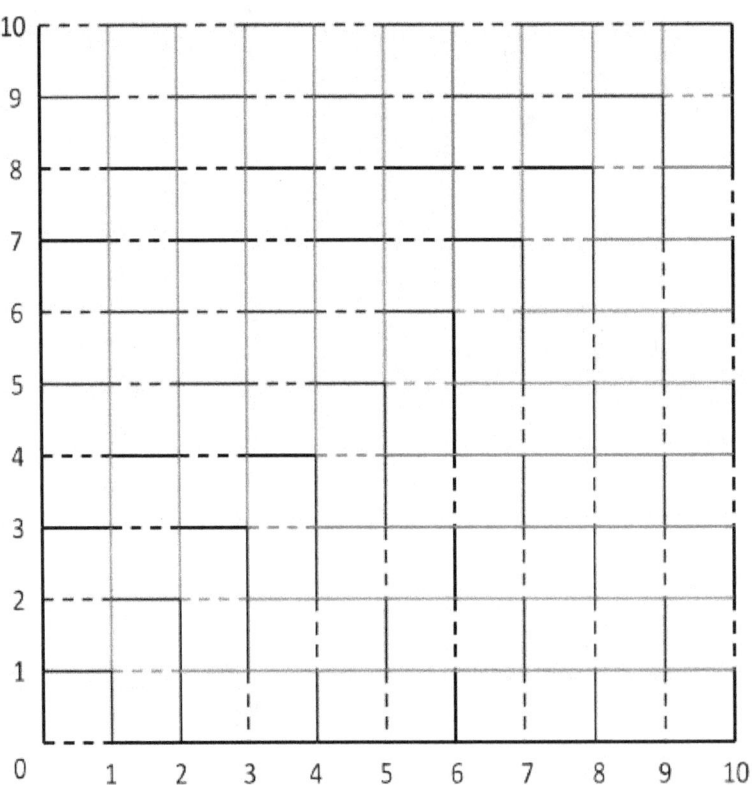

From Table 3 and Figure 2, it is clear that the upper bound of $P(n)$ is given by the Triangular Numbers,

$$P(n) \leq T(n) = n\,(n+1)\,/\,2.$$

But there are cases when $P(n) < T(n)$ as for $n = 4$, $T(4) = 10$ but $P(4) \leq 9$ as demonstrated in Figure 3.

Figure 3
4 × 4 Grid

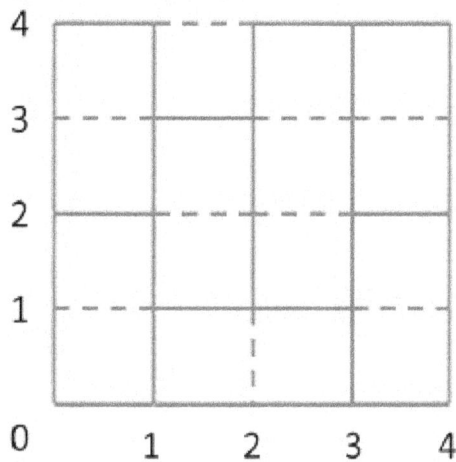

It is interesting to note that

$$0 \leq P(n)\,/\,E(n) \leq T(n)\,/\,E(n) = 0.25$$

where $E(n) = 2n\,(n+1)$ denotes the number of edges in $n \times n$ grid.

Conclusions

The second algorithm provides sharper upper bounds of $P(n)$ and consumes less time. That it consumes less time is explainable as the search is on a smaller number of edges. However the fact that it provides sharper bounds are worth investigating.

The problem could be extended to any grid of regular shape and to higher dimensions as well.

31

References

Vanishing Squares, http://ken.duisenberg.com/potw/archive/arch99/990906sol.html

About the Second Droz-Farny Circle

Ion Pătraşcu

Colegiul National Fratii Buzeşti

Craiova, România

Florentin Smarandache

Universitatea, New Mexico

Abstract

In this article, we prove the theorem relative to the *second Droz-Farny circle*, and a sentence that generalizes it.

Mihalescu [1] states that the following theorem is due to J. Neuberg (*Mathesis*, 1911).

First Theorem. The circles with its centers in the middles of triangle ABC passing through its orthocenter H intersect the sides BC, CA and AB respectively in the points A_1, A_2, B_1, B_2 and C_1, C_2, situated on a concentric circle with the circle circumscribed to the triangle ABC (**the second Droz-Farny circle**).

Proof. Referring to figure 1, we denote by M_1, M_2, M_3 the middles of ABC triangle's sides. Because $AH \perp M_2 M_3$ and H belongs to the circles with centers in M_2 and M_3, it follows that AH is the radical axis of these circles, therefore we have $AC_1 \cdot AC_2 = AB_2 \cdot AB_1$. This relation shows that B_1, B_2, C_1, C_2 are concyclic points, because the center of the circle on which they are situated is O, the center of the circle circumscribed to the triangle ABC. Hence we have that:

$$OB_1 = OC_1 = OC_2 = OB_2 \quad (1).$$

Figure 1

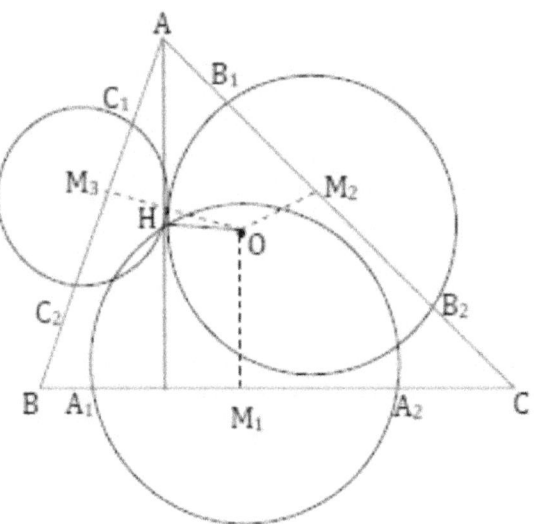

Analogously, O is the center of the circle on which the points A_1, A_2, C_1, C_2 are situated. Therefore,

$$OA_1 = OC_1 = OC_2 = OA_2 \quad (2).$$

Furthermore, O is the center of the circle on which the points A_1, A_2, B_1, B_2 are situated, and so:

$$OA_1 = OB_1 = OB_2 = OA_2 \quad (3).$$

The relations (1), (2), (3) show that the points $A_1, A_2, B_1, B_2, C_1, C_2$ are situated on a circle having the center in O, called *the second Droz-Farny circle*.

Proposition. The radius of the second Droz-Farny circle is given by:

$$R_2^2 = 5e^2 - \frac{1}{2}(a^2 + b^2 + c^2).$$

Proof. From the right triangle OM_1A_1, using the Pythagorean theorem, it follows that:

$$OA_1^2 = OM_1^2 + A_1M_1^2 = OM_1^2 + M_1M_2.$$

From the triangle BHC, using the median theorem, we have:

$$HM_1^2 = \frac{1}{4}[2(BH^2 + CH^2) - BC^2].$$

However, in a triangle, $AH = 2OM_1$, $BH = 2OM_2$, $CH = 2OM_3$, hence:

$$HM_1^2 = 2OM_2^2 + 2OM_3^2 = \frac{a^2}{4}.$$

But $OM_1^2 = R^2 - \frac{a^2}{4}$; $OM_2^2 = R^2 - \frac{b^2}{4}$; $OM_3^2 = R^2 - \frac{c^2}{4}$, where R is the radius of the circle circumscribed to the triangle ABC.

Therefore, we have that $OA_1^2 = R_2^2 = 5R^2 - \frac{1}{2}(a^2 + b^2 + c^2).$

Remarks

1. We can compute $OM_1^2 + M_1M_2$ using the median theorem in the triangle OM_1H for the median M_1O_9 (O_9 is the center of the nine points circle, i.e. the middle of (OH)). Because $O_9M_1 = \frac{1}{2}R$, we obtain: $R_2^2 = \frac{1}{2}(OM^2 + R^2)$. In this way, we can prove the *Theorem* computing OB_1^2 and OC_1^2.

2. The statement of the *First Theorem* was the subject no. 1 of the 49th International Olympiad in Mathematics, held at Madrid in 2008.

3. The *First Theorem* can be proved in the same way for an obtuse triangle, but it is obvious that for a right triangle, the second Droz-Farny circle coincides with the circle circumscribed to the triangle ABC.

4. The *First Theorem* appears as proposed problem in [2].

Second Theorem. The three pairs of points determined by the intersections of each circle with the center in the middle of triangle's side with the respective side are on a circle if and only these circles have as radical center the triangle's orthocenter.

Proof. Let M_1, M_2, M_3 the middles of the sides of triangle ABC and let $A_1, A_2, B_1, B_2, C_1, C_2$ the intersections with BC, CA, AB respectively of the circles with centers in M_1, M_2, M_3.

Let us suppose that $A_1, A_2, B_1, B_2, C_1, C_2$ are concyclic points. The circle on which they are situated has evidently the center in O, the center of the circle circumscribed to the triangle ABC. The radical axis of the circles with centers M_2, M_3 will be perpendicular on the line of centers M_2M_3, and because A has equal powers in relation to these circles, since $AB_1 \cdot AB_2 = AC_1 \cdot AC_2$, it follows that the radical axis will be the perpendicular taken from A on M_2M_3, i.e. the height from A of triangle ABC.

Furthermore, it ensues that the radical axis of the circles with centers in M_1 and M_2 is the height from B of triangle ABC and consequently the intersection of the heights, hence the orthocenter H of the triangle ABC is the radical center of the three circles.

Reciprocal. If the circles having the centers in M_1, M_2, M_3 have the orthocenter with the radical center, it follows that the point A, being situated on the height from A which is the radical axis of the circles of centers M_2, M_3 will have equal powers in relation to these circles and, consequently, $AB_1 \cdot AB_2 = AC_1 \cdot AC_2$, a relation that implies that B_1, B_2, C_1, C_2 are concyclic points, and the circle on which these points are situated has O as its center. Similarly, $BA_1 \cdot BA_2 = BC_1 \cdot BC_2$, therefore A_1, A_2, C_1, C_2 are concyclic points on a circle of center O. Having $OB_1 = OB_2 = OC_1 = OC_2$ and $OA_1 \cdot OA_2 = OC_1 \cdot OC_2$, we get that the points $A_1, A_2, B_1, B_2, C_1, C_2$ are situated on a circle of center O.

Remarks.

1. The *First Theorem* is a particular case of the *Second Theorem*, because the three circles of centers M_1, M_2, M_3 pass through H, which means that H is their radical center.

2. The Problem 525 from [3] drives us to the following *Proposition* that provides the way to construct the circles of centers M_1, M_2, M_3 that intersect the sides in points that belong to a Droz-Farny circle of type 2.

Proposition. Referring to figure 2, the circles $C\left(M_1, \frac{1}{2}\sqrt{k + a^2}\right)$, $C\left(M_2, \frac{1}{2}\sqrt{k + b^2}\right)$, $C\left(M_3, \frac{1}{2}\sqrt{k + c^2}\right)$ intersect the sides BC, CA, AB respectively in six concyclic points; k is a conveniently chosen constant, and a, b, c are the lengths of the sides of triangle ABC.

Proof. According to the *Second Theorem*, it is necessary to prove that the orthocenter H of triangle ABC is the radical center for the circles from the hypothesis.

The power of H in relation with $C\left(M_1, \frac{1}{2}\sqrt{k + a^2}\right)$ is equal to $HM_1^2 - \frac{1}{4}(k + a^2)$. We observed that $M_1^2 = 4R^2 - \frac{b^2}{2} - \frac{c^2}{2} - \frac{a^2}{4}$, therefore $HM_1^2 - \frac{1}{4}(k + a^2) = 4R^2 - \frac{a^2+b^2+c^2}{4} - \frac{1}{4}k$. We use the same expression for the power of H in relation to the circles of centers M_2, M_3, hence H is the radical center of these three circles.

Figure 2

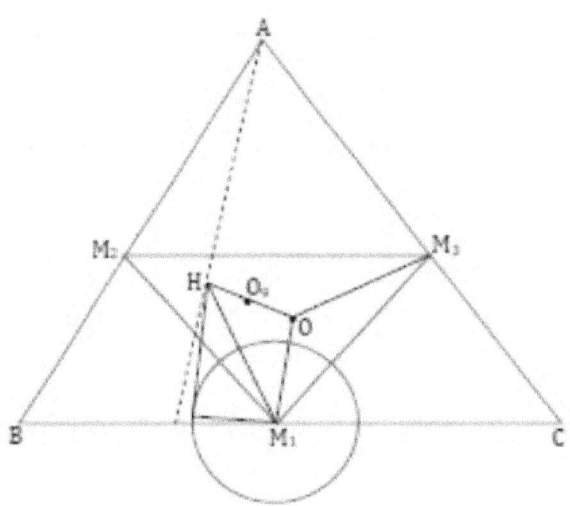

Bibliography

[1] C. Mihalescu: *Geometria elementelor remarcabile [The Geometry of Outstanding Elements]*. Bucharest: Editura Tehnică, 1957.

[2] I. Pătraşcu: *Probleme de geometrie plană [Some Problems of Plane Geometry]*, Craiova: Editura Cardinal, 1996.

[3] C. Coşniţă: *Teoreme şi probleme alese de matematică [Theorems and Problems]*, Bucureşti: Editura de Stat Didactică şi Pedagogică, 1958.

[4] I. Pătraşcu, F. Smarandache: *Variance on Topics of Plane Geometry*, Educational Publishing, Columbus, Ohio, SUA, 2013.

Tilings of Quadrants by *L*-ominoes and Notched Rectangles

Aaron Calderon

University of Nebraska-Lincoln

aaron.calderon256@gmail.com

Samantha Fairchild

Houghton College

samantha.fairchild15@houghton.edu

Viorel Nitica

West Chester University

vnitica@wcupa.edu

Samuel Simon

Carnegie Mellon University

slsimon@cmu.edu

Abstract

In this paper, we examine tilings of the four quadrants in a Cartesian coordinate system by tile sets consisting of *L*-shaped polyominoes and notched rectangles. We start with tile sets consisting of an *L*-shaped polyomino and a notched rectangle, appearing from the dissection of a $n \times n$, $n \geq 3$, square, and of the symmetries of these two tiles about the first diagonal. In this case, a tiling of a quadrant is said to follow the rectangular pattern if it reduces to a tiling by $n \times n$ squares, each of the squares in turn tiled by an *L*-shaped polyomino and a notched rectangle. We show that for every tile sets as above, with the possible exception of one appearing from the dissection of a 3×3 square, there exists at least a quadrant for which any tiling has to follow the rectangular pattern. We further consider tilings of the quadrants with tile sets appearing from similar dissections of $mn \times n$, $n \geq 3$, $m \geq 2$, rectangles and show that for every one of them there exists at least a quadrant for which every tiling has to follow the rectangular pattern. Our results have consequences for tilings of other regions in plane, in particular rectangles and infinite half strips. A rectangle can be tiled by the above tile sets if and only if both sides are divisible by n and one side is divisible by mn, and an infinite half strip can be tiled if and only if its width is divisible by n. In all of the above cases, the rectangular pattern of a tiling persists if we add an extra n × n square to the tile set. Our technique of proof is to use induction along a staircase line built out of n × n squares and to show that the existence of a tile in an irregular position propagates towards the edges of the quadrant, and eventually leads to a contradiction. Further, we look at similar tile sets appearing from dissections of rectangles m × n of coprime sides m, n n ≥ 2. Here there exist tilings that do not follow the rectangular pattern.

1. Introduction

In this article we study tiling problems for regions in a square lattice by certain sets of tiles consisting of polyominoes. Polyominoes were introduced by Solomon W. Golomb [4] and the standard reference about tiling with polyominoes is the book *Polyominoes* [8]. The study of polyominoes has generated a great amount of mathematical literature. We continue here the investigation started in the papers [2,10], where the notion of *rectangular pattern* is introduced for a tiling of the first quadrant by tile sets generated by the dissection of a rectangle in two L-shaped polyominoes and their symmetries about the first diagonal.

The 1×1 squares in the square lattice are called *cells*. For an $a \times b$ rectangle, a is the height and b is the base. We consider the four possible dissections of a $k \times n$ rectangle, $3 \leq n \leq k$, into two parts: an L-shaped polyomino of width 1 that extends along the height of the dissected rectangle, and a remaining notched rectangle. The missing part from the notched rectangle is always a cell, as exhibited in Figure 1, where $n = 3, k = 6$. We denote these dissections by $C_1, C_2,\ C_3$ and C_4.

Figure 1

Dissections C_1, C_2, C_3, C_4

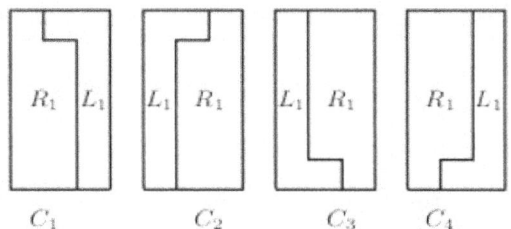

Each dissection determines two tiles from the corresponding tile set. To complete the tiling set to a total of four tiles, we reflect the pieces resulting from the dissection about the first bisector $y = x$. Only translations, that is, no rotations or reflections, of the tiles in the tile set are allowed in a tiling. The L-shaped polyominoes are denoted by L_1 and their reflections by L_2 The notched rectangles are denoted by R_1 and their reflections by R_2. We refer to L_1, R_1 as the vertical tiles and to L_2, R_2 as the horizontal tiles. The tile set for C_1 is shown in Figure 2, a). We denote by $T(C_i, k, n)$ the tile set given by the dissection $C_i, 1 \leq i \leq 4$. If an extra $n \times n$ square, which we will denote by S, is added to the tile set, we denote the tile set $T^+(C_i, k, n)$. A $T^+(C_1, k, n)$ tile set is shown in Figure 2, b).

Figure 2

Tiling Sets

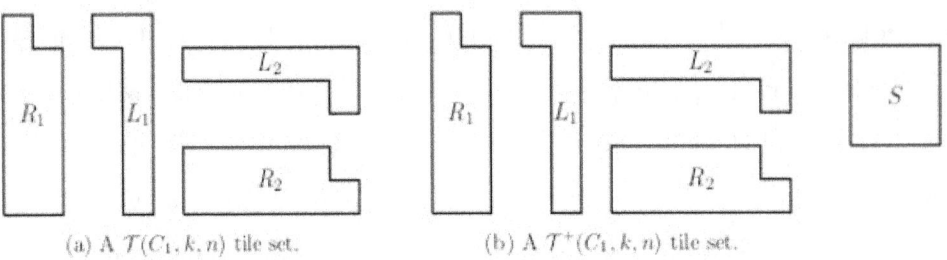

(a) A $\mathcal{T}(C_1, k, n)$ tile set.　　　　(b) A $\mathcal{T}^+(C_1, k, n)$ tile set.

Definition: A tiling by $T(C_i, k, n)$, $1 \le i \le 4$, $3 \le n \le k$, of a region in the plane is said to follow the rectangular pattern if it reduces to a tiling by $k \times n$ and $n \times k$ rectangles, each tiled in turn by two pieces from the tile set. A tiling by $T^+(C_i, k, n)$, $1 \le i \le 4$, $3 \le n \le k$, of a region in the plane is said to follow the rectangular pattern if it reduces to a tiling by $n \times n$ squares and by $k \times n$ and $n \times k$ rectangles, each rectangle tiled in turn by pieces from the tile set. If all tilings of a quadrant Q by a tile set T or T^+ follow the rectangular pattern, we say that T, respectively T^+, is rigid with respect to Q.

For simplicity, in the future we will denote the four quadrants in a Cartesian coordinate system respectively by Q_1, Q_2, Q_3, Q_4. Figure 3 shows a tiling of Q_1 by $T(C_2, 3, 3)$ that follows the rectangular pattern. The general problem of interest to us is the following:

Problem 1. Identify which of the tile sets $T(C_i, k, n)$, $T^+(C_i, k, n)$, $1 \le i \le 4$, $3 \le n \le k$, are rigid with respect to the quadrants Q_1, Q_2, Q_3, Q_4.

Figure 3

A Rigidity Problem

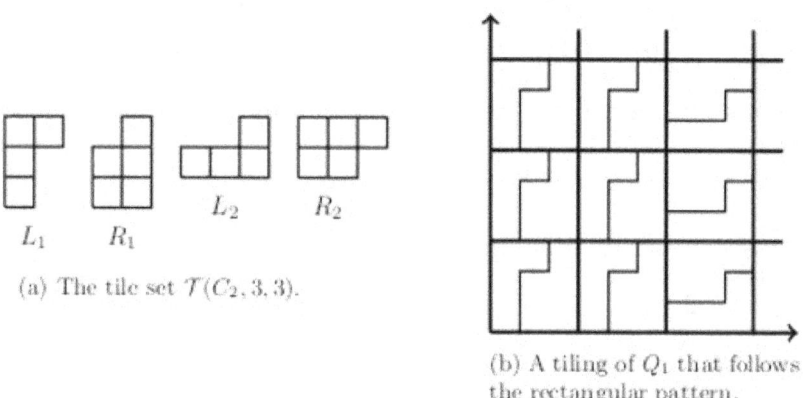

(a) The tile set $\mathcal{T}(C_2, 3, 3)$.

(b) A tiling of Q_1 that follows the rectangular pattern.

The study of rigid tile sets with respect to Q_1 was started in [2], inspired by a problem from recreational mathematics [5,9]. We were interested if the skewed tetromino introduced by Golomb in [5] is a replicating tile of odd order. The tile set discussed in [2] appears from the dissection of a 4×2 rectangle in two L-tetrominoes and is made of all four L-shaped ribbon tetrominoes. See Figure 4. It is shown in [2] that this tile set is rigid with respect to Q_1.

Figure 4

The Four L-Shaped Ribbon Tetrominoes

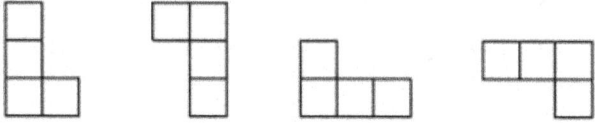

Further progress is done in the paper [10], where the tile sets are defined by dissections of rectangles of base of length 2 into two L-shaped polyominoes, not necessarily congruent. The tile sets consist of four tiles: the two L-shapes generated by the dissection and their symmetrics about the first bisector. When the height of the dissected rectangle is odd, the problem is completely solved by showing that always there exist tilings by the tile set that do not follow the rectangular pattern. The situation is more complex when the height of the dissected rectangle is even. If the tiles appearing from the dissection are congruent, the rigidity problem is completely solved. We identify both rigid and nonrigid tiling sets and they appear in infinite families. If the tiles appearing from the dissection are not congruent, several cases are solved, identifying both

43

rigid and nonrigid tiling sets, and several cases are left open. The discussion above explains the approach taken in this paper. While it is desirable to completely solve the rigidity problem for tile sets given by a dissection of a rectangle in two polyominoes, it is difficult to do so even in the case of a rectangle of base of length 2. This is why we only study special classes of rectangles/dissections that naturally generalize previous work. Our main results are listed below.

Theorem 1.

1) The tile set $T^+(C_1, mn, n), n \geq 3, m \geq 1$, is rigid with respect to Q_1.

2) The tile set $T(C_1, mn, n), n \geq 3, m \geq 1$, is not rigid with respect to Q_2, Q_3, Q_4.

3) The tile set $T^+(C_3, mn, n), n \geq 3, m \geq 1$, is rigid with respect to Q_3.

4) The tile set $T(C_3, mn, n), n \geq 3, m \geq 1$, is not rigid with respect to Q_1, Q_2, Q_4

5) The tile set $T^+(C_2, mn, n), n \geq 3, m > 1$, is rigid with respect to Q_2.

6) The tile set $T(C_2, 3,3)$ is not rigid with respect to Q_2, Q_3, Q_4.

7) The tile set $T^+(C_2, n, n), n \geq 4$, is rigid with respect to all quadrants.

8) The tile set $T^+(C_4, mn, n), n \geq 3, m > 1$, is rigid with respect to Q_4.

9) The tile set $T(C_4, 3,3)$ is not rigid with respect to the Q_1, Q_2, Q_4.

10) The tile set $T^+(C_4, n, n), n \geq 4$, is rigid with respect to all quadrants.

We observe that positive rigidity results about $T^+(C_i, k, n)$ imply similar positive rigidity results about $T(C_i, k, n)$ and negative rigidity results about $T(C_i, k, n)$ imply similar negative rigidity results for $T^+(C_i, k, n)$. The theorem leaves open the following problem.

Problem 1. Decide if the following tile sets are rigid:

1) $T(C_2, 3,3)$ with respect to Q_1.

2) $T(C_2, mn, n), n \geq 3, m > 1$, with respect to Q_1, Q_3, Q_4.

Several cases from Theorem 1 follow from others due to symmetries. The tile set $T(C_3, k, n)$ can be obtained from $T(C_1, k, n)$ via a rotation by $180°$, and the tile set $T(C_4, k, n)$ can be obtained from $T(C_2, k, n)$ via a rotation by $180°$. Moreover, $T(C_2, k, n)$ can be obtained from $T(C_4, k, n)$ via a symmetry about the second bisector $y = -x$. Thus in order to prove Theorem 1 we only need to show the cases 1), 2), 5), 6) and 7).

44

Our technique of proof is to use induction along a staircase line built out of $n \times n$ squares and to show that the existence of a tile in an irregular position propagates towards the edges of the quadrant, and eventually leads to a contradiction. A similar approach is used in [10], even though certain combinatorial objects used in the proof ("the gap", "the tile in an irregular position", "the new staircase",) are different. The presence of a coloring invariant allows for a simpler proof in [2]. In contrast, neither of the results in Theorem 1 can be obtained from coloring invariants. It is well known that the most general coloring argument reduces to the existence of a signed tiling. Our next result shows that any region in the square lattice can be signed tiled by our tile sets, so no obstructions may appear in this way.

Theorem 2. All tile sets considered in Theorem 1 can sign tile any region in the square lattice.

Theorem 3. If $3 \leq n \leq k$ are coprime, all quadrants have tilings by $T(C_i, k, n)$, $1 \leq i \leq 4$, that do not follow the rectangular pattern.

Our results suggest some conjectures for the problem of tiling Q_1, Q_2, Q_3, Q_4 by $T(C_i, k, n)$:

Conjecture 1. If the height of the dissected rectangle is a multiple of the base $n \geq 2$, then both rigid and nonrigid tilings are possible for infinite families of rectangles/dissections.

Conjecture 2. If the sides of the dissected rectangle have a proper common factor greater then 1, and no side is a multiple of the other, the problem is completely open. The first case of interest appears from the dissection of a 6×4 rectangle.

Solving the rigidity problem for quadrants has applications to tiling other regions in the plane such as rectangles and half-infinite strips. See [2,10] for the case $n = 2$. The proofs of Corollaries 1, 2, 3 below are similar to those in [2].

Corollary 1. With a possible exception for $T^+(C_2, 3, 3)$, every tiling of a rectangle by one of the tiling sets $T^+(C_i, mn, n)$, $1 \leq i \leq 4$, $n \geq 3$, $m \geq 1$, follows the rectangular pattern. Consequently, with a possible exception for $T^+(C_2, 3, 3)$, a rectangle can be tiled by $T^+(C_i, mn, n)$, $1 \leq i \leq 4$, $n \geq 3$, $m \geq 1$, if and only if both of its sides are divisible by n.

The question of what rectangles can be tiled by $T(C_i, mn, n)$, $1 \leq i \leq 4$, $n \geq 3$, $m > 1$, is reduced, via Theorem 1, to the question of what rectangles can be tiled by $mn \times n$ and $n \times mn$ rectangles. Any such rectangle has both sides divisible by n. We recall that it was proved by de Bruin [1] that an $a \times b$ rectangle can be tiled by $k \times 1$ and $1 \times k$ bars if and only if k divides one of the sides of the rectangle. This implies:

Corollary 2. With a possible exception for $T(C_2, 3,3)$, every tiling of a rectangle by one of the tiling sets $T(C_i, mn, n)$, $1 \leq i \leq 4$, $n \geq 3$, $m \geq 1$, follows the rectangular pattern. Consequently, with a possible exception for $T(C_2, 3,3)$, a rectangle can be tiled by $T(C_i, mn, n)$, $1 \leq i \leq 4$, $n \geq 3$, $m \geq 1$, if and only if one of its sides is divisible by n and the other is divisible by mn.

Corollary 3. With a possible exception for $T(C_2, 3,3)$, a half-infinite strip of width not divisible by n cannot be tiled by $T(C_i, mn, n)$, $1 \leq i \leq 4$, $n \geq 3$, $m \geq 1$.

Problem 3. Show that a double infinite strip of width not divisible by n cannot be tiled by $T(C_i, mn, n)$, $1 \leq i \leq 4$, $n \geq 3$, $m \geq 1$.

Let us observe that certain computational effort was made in the investigation of tiling problems with tile sets consisting of a single tile, see the papers by Golomb [6], Hochberg [3], and Reid [12]. This paper, [10] and [2] present a new approach for the study of tile sets with several tiles, a study which was under looked in the literature. Among the few papers in this direction we mention those of Golomb [7], Pak [11] and Sallows [13,14].

The rest of the paper is organized as follows: In Section 2 we show the proof of the rigidity result for $T^+(C_1, n, n)$, $n \geq 3$ and start the proof of Theorem 1, 1). We do this in order to outline the main idea of the proof in the general case. In Section 3 we show the proof of the rigidity result for $T^+(C_2, n, n)$, $n \geq 4$. In Section 4 we show the proof of the rigidity result for $T^+(C_1, mn, n)$, $n \geq 3$, $m \geq 2$, finishing the proof of Theorem 1, 1). In Section 5 we show the proof of the rigidity result for $T^+(C_2, mn, n)$, $n \geq 3$, $m \geq 2$, proving Theorem 1, 5). In Section 6 we show the nonrigid results for $T(C_2, mn, n)$, $m \geq 2$, Theorem 1, 2). In Section 7 we show the nonrigid results for $T(C_3, 3,3)$, Theorem 1, 6). Theorem 3 is proved in Section 8 and Theorem 2 is proved in Section 9.

2. Tiling Q_1 by T^+ (C_1, n,n).

For the rest of the paper, in the proofs of the positive rigidity results, we use an additional $n \times n$ square lattice which has the same origin and coordinate axes as the initial 1×1 square lattice. The squares in this lattice are called n-squares. Note that the vertices of an n-square have the coordinates of all vertices divisible by n. For a fixed quadrant Q and a fixed tile set T, if a tiling of Q is given, an n-square is said to be *regularly covered* if it is completely covered by an $n \times n$, $mn \times n$, or $n \times mn$ rectangle with the coordinates of all vertices divisible by n, rectangle which in turn is tiled by T. Our proofs are by induction on the k-th iteration of the $n \times n$ staircase line in Q. The picture of the staircase for Q_1 is shown in Figure 18, b), and for Q_2 is shown in Figure 22, b). When we refer to a staircase line, here and in the future sections, we assume that all n-squares belonging to the finite region bounded by the staircase and the coordinate axes are already regularly covered.

In this section we prove the rigidity of the tile set $T^+(C_1, n, n), n \geq 3$ with respect to Q_1, thus proving case 1), $m = 1$, in Theorem 1. During the proof, for simplicity, we draw the figures mostly for $n = 3$, but the arguments are general.

Consider the base case, $k = 1$. If the corner n-square is not regularly covered, then the corner cell can only be covered by R_1 or R_2. Due to the symmetry, we may assume that we use R_1. Consider cell 1 in Figure 5, a), the notch of R_1 tile. If it is irregularly covered by L_1 or R_1, then the gray regions in Figures 5, c), d), cannot be completely covered by the tile set. This is clear from the figures if $n = 3$. If $n > 3$, only vertical L_1 tiles can be introduced in the gray region and they leave certain cells uncovered.

If cell 1 is irregularly covered by L_2 or R_2, then the gray regions in Figures 5, e), f), cannot be completely covered by the tile set. This is clear from the figures if n = 3. If n > 3, no tile can cover the cells in the gray region in Figure 5, e), adjacent to the L_2 tile, and only vertical L_1 tiles can be introduced in the gray region in Figure 5, f), and they leave certain cells uncovered.

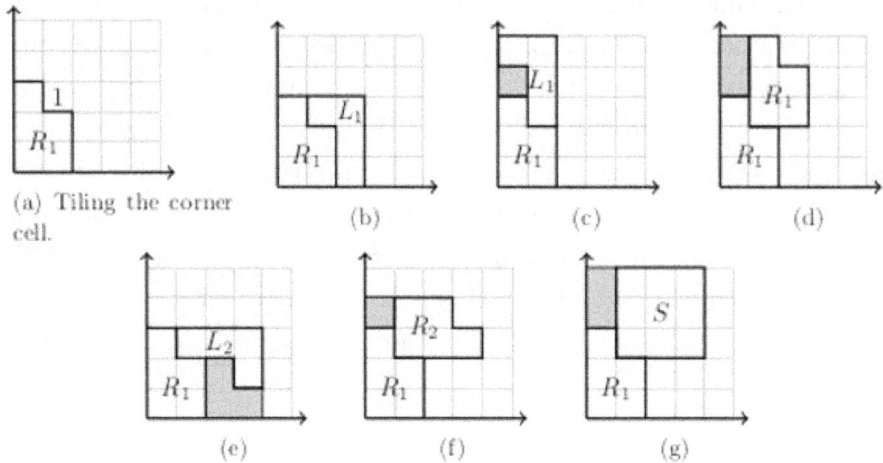

(a) Tiling the corner cell.   (b)   (c)   (d)

(e)   (f)   (g)

If cell 1 is covered irregularly by S, then the gray region in Figure 5, g), cannot be completely covered by the tile set. If n > 3, only vertical L$_1$ tiles can be introduced in the gray region and they leave certain cells uncovered. We also observe that the short end of L$_2$ cannot cover the notch in R$_1$. Hence the only left possibility, shown in Figure 5, b), leads to a regular tiling of the corner n-square. In the future we will keep the discussion of the cases n > 3 quite brief due to the fact that the reasoning is usually close to the arguments above.

Now consider the staircase line for $k > 1$. Call the rightmost n-square above the staircase that is not regularly covered X_i (see Figure 6, a)). Look at what tile covers the bottom left cell of X_i. It is R_1 or R_2. The n-square X_{i-1} to the lower right of X_i is regularly covered by our choice of X_i, so the covering of the corner cell by R_2 forces a regular covering of X_i via an argument similar to that used in the base case. We provide more details only for the case shown in Figure 5, e), see Figure 6, b). If $n \geq 3$, the cell in the gray region immediately above the R_2 tile cannot be covered by any tile in the tile set.

Assume that the corner cell is covered by R$_1$, as in Figure 6, c). If we cover cell 1 (the notch) by the short end of L$_1$ we arrive at a regular covering of X$_1$. We consider other possible coverings of cell 1 and show that they lead to a contradiction.

Figure 6

Tiling the n-Square X_i in Q_1 by T^+ (C_1, n, n), n ≥ 3

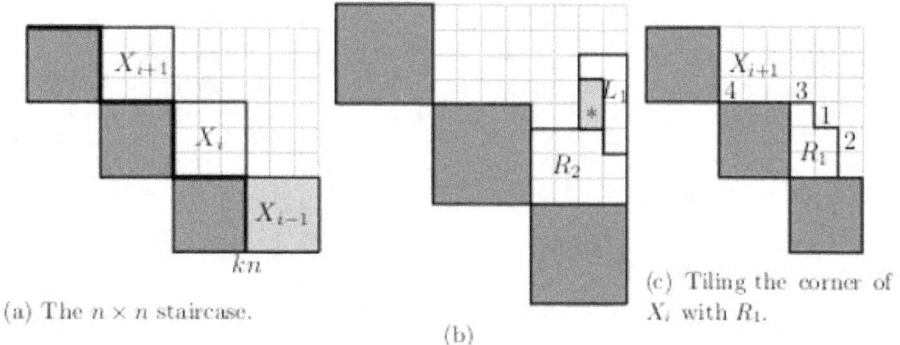

(a) The $n \times n$ staircase.

(b)

(c) Tiling the corner of X_i with R_1.

Case 1: We use the long end of L_1 to tile cell 1. The possible tilings of cell 3, diagonally adjacent to cell 1, may use R_2, which leaves an untileable region of width 1 on the left side of the R_2 tile, or, if $n \geq 4$, may use L_2, which leaves an untilable region below the L_2 tile. See the light gray regions in Figures 7, a), b).

Case 2: If cell 1 is covered with the long end of L_2, cell 2, diagonally adjacent to cell 1, cannot be tiled. See Figure 7, c).

Case 3: Suppose cell 1 is covered with the short end of L_2, as in Figure 8. To tile cell 4 we are forced to use R_1. We consider which tiles can be used to tile cell 5, the notch of an R_1 tile.

Subcase 1: If we to use R_1, R_2, S or the long end of L_2, then we have an untileable region containing cell 6, diagonally adjacent to cell 5, which is bounded above by the new tile and below by the L_2 tile that covers cell 1.

Subcase 2: We use the long end of L_1. Then we reach the contradiction in Case 1.

Subcase 3: Therefore, the only way to tile cell 5 which does not lead to a contradiction right away is by L_2. Continuing in this fashion, a new staircase above the initial one develops. We eventually reach the y-axis, see Figure 8, b). Now, any attempt to tile cell 7, the notch of an R_1 tile, results in a region containing the cell diagonally adjacent to cell 7, labeled 8, that cannot be covered.

Figure 7

Tiling the Corner of X_i With R_1

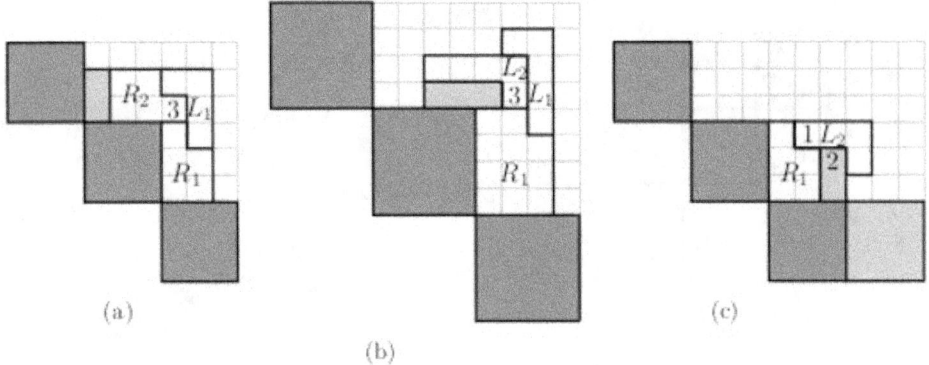

(a)

(b)

(c)

Figure 8

Tiling Cell 1 by L_2 and the New Staircase

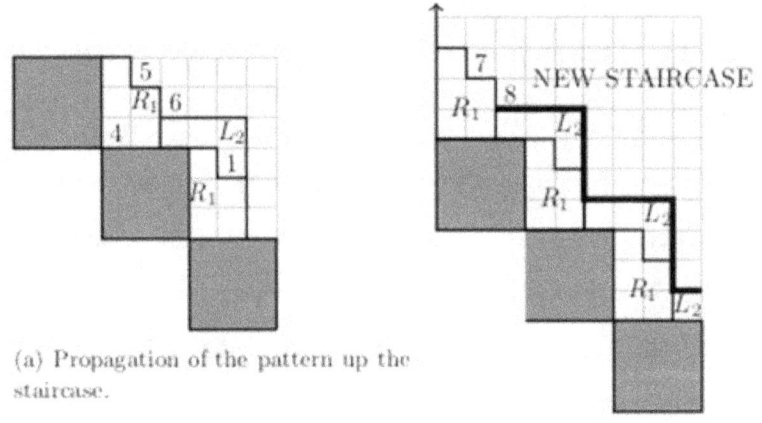

(a) Propagation of the pattern up the staircase.

(b) The end of the propagation.

Case 4: We now tile cell 1 with R_1, R_2 or S, see Figure 9. We may tile cell 4 in the lower left corner of X_{i+1} only with R_1, R_2 or S. If we use R_2, cell 3 can be covered only by L_1, which forces the notch from the R_2 tile to be untilable. See Figure 9, a). If we use S, cell 3 cannot be covered by the tile set. See Figure 9, b). Thus we have to tile cell 4 with R_1. Now the only way we can tile cell 3 and the cell on its immediate left is with R_1, as in Figure 9, c). We look now at cell 5.

50

Subcase 1: If we tile cell 5 by L_2, then we reach the pattern of Case 3.

Subcase 2: Any attempt to tile cell 5 with L_1 forces R_2 to tile cell 10. Then we reach a contradiction as in Case 1.

Figure 9

Tiling Cell 1 by R₁, R₂ or S

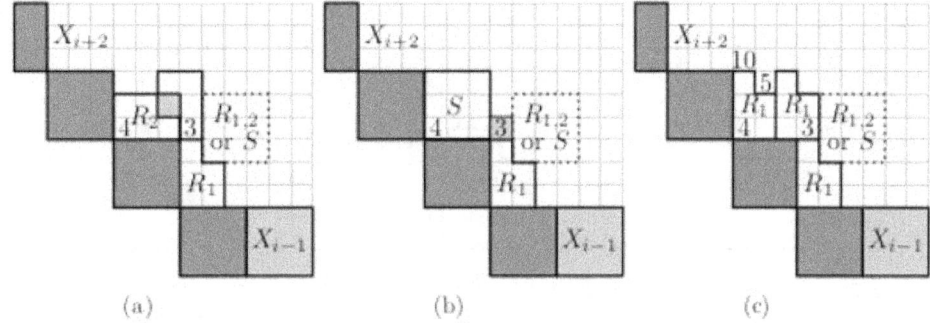

(a)   (b)   (c)

The summary of the previous discussion is that, if X_i is not regularly covered , then either we reach a contradiction right away, or we reach a contradiction after placing several L_2 tiles along the new staircase.

3. Tiling of the quadrants by $T^+(C_2,n,n)$.

In this section we prove the rigidity of the tile set $T^+(C_2, n, n), n \geq 4$, with respect to all quadrants, thus proving Theorem 1, case 7). Due to the invariance of $T^+(C_2, n, n), n \geq 4$, under a reflection in the first bisector, it is enough to show the proofs for Q_1, Q_2, Q_3. We will draw the pictures for $n = 4$, but our arguments are general.

The case of Q_2. Consider the base case $k = 1$. The corner cell of Q_2 can only be covered with R_1 or L_2. If we use R_1, then any tile other than L_2 that is used to cover cell 1, the notch of the R_1 tile, makes it impossible to tile the gray regions in Figure 10. Similarly, if we use L_2, any tile used to cover cell 1 in Figure 11 makes it impossible to tile the gray regions in Figure 11. Thus the corner n-square must be regularly covered.

51

Figure 10

Tiling the Corner n-Square in Q_2 by $T^+(C_2, n, n)$, $n \geq 3$, Part 1

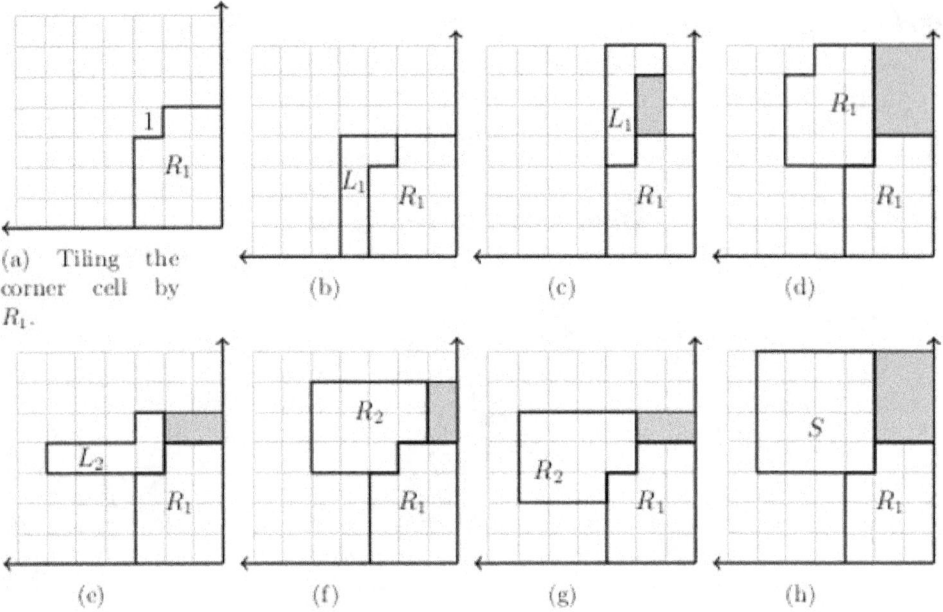

(a) Tiling the corner cell by R_1.

(b)

(c)

(d)

(e)

(f)

(g)

(h)

Figure 11

Tiling the Corner n-Square in Q₂ by T⁺(C₂, n, n), n ≥ 3, Part 2

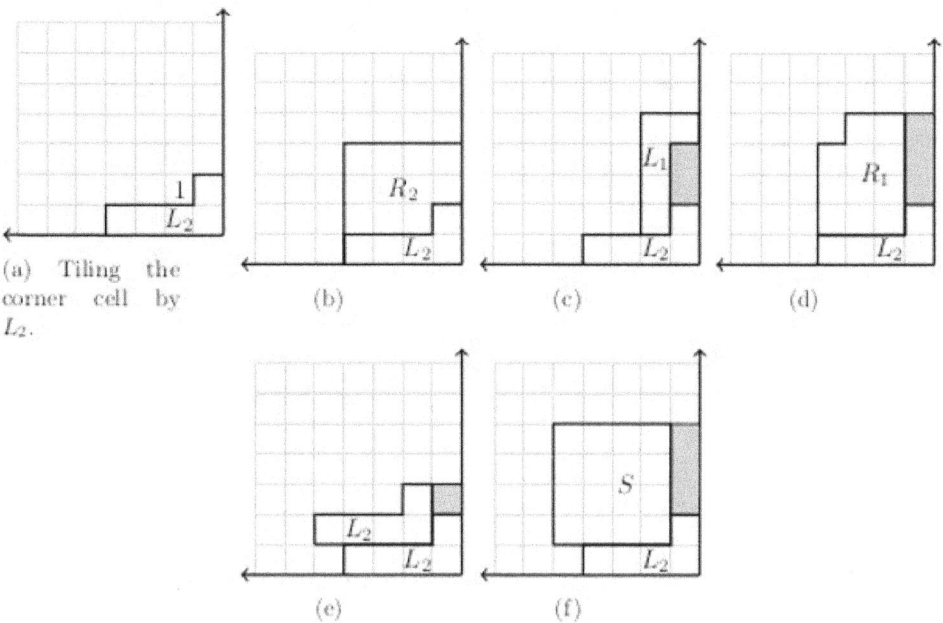

(a) Tiling the corner cell by L_2.

(b)

(c)

(d)

(e)

(f)

Now let $k > 1$. Call X_i the bottom most n-square above the staircase that does not follow the rigid pattern. By our assumption, the n-square X_{i-1} to the top right of X_i is regularly covered. We look at what tile covers the bottom right corner cell of X_i (Figure 12). It can be only R_1 or L_2. If we use R_1 and the n-square above the R_1 tile is regularly covered there is no way to cover the notch of the R_1 tile. Otherwise, due to the fact that the n-square X_{i-1} is regularly covered, in either case we can conclude as in the base case $k = 1$. So X_i is regularly covered, and $T^+(C_2, n, n)$, $n \geq 4$, is rigid with respect to Q_2.

Figure 12

Tiling the n-Square X_i in Q_2 by $T^+(C_2, n, n)$, $n \geq 3$

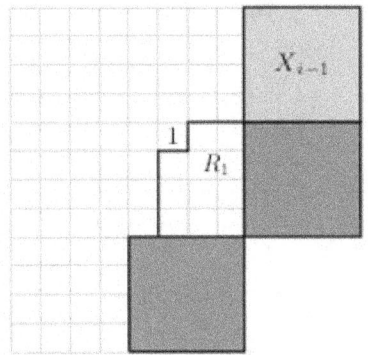

(a) Tiling the corner of X_i with R_1.

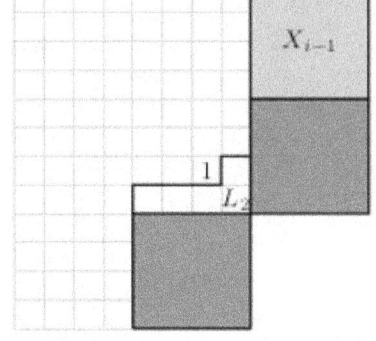

(b) Tiling the corner of X_i with L_2.

The case of Q_1. Consider the base case $k = 1$. If the corner cell is covered by R_1, the notch can only be covered by L_1. See Figure 13. Then we are left with a gray region to the right of the L_1 tile that cannot be covered completely by tiles in $T^+(C_2, n, n)$. Similarly, by symmetry, we may not cover the corner cell by R_2. If we use L_1 to tile the corner cell, then we have a $(n - 1) \times 1$ region adjacent to the tile L_1 which can only be covered without contradiction by R_2. But then we are forced to tile the notch of the R_2 tile by L_2, creating a gray region that cannot be tiled. See Figure 13. Therefore the bottom left corner must be regularly covered.

Figure 13

Tiling the Corner n-Square in Q₁ by T⁺(C₂, n, n), n ≥ 3

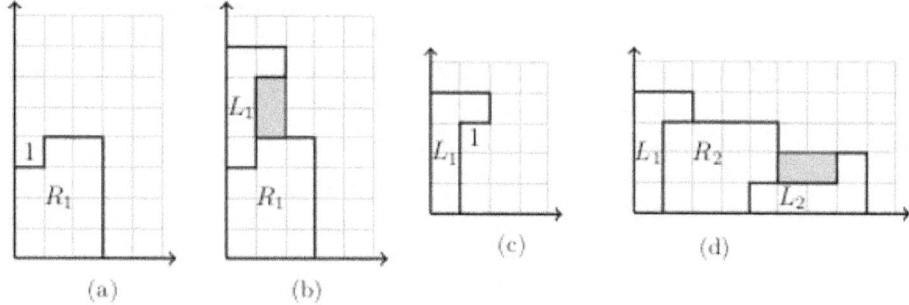

Now consider $k > 1$. Call the furthest right irregularly covered n-square X_i. If we tile its left lower corner cell with L_1, as X_{i-1} is regularly covered, the same argument as in base case apply, leading to a regular tiling of X_1. Tiling the cell with R_1 or R_2 leaves the notch which can only be covered as in the corner cases. We are then left with the same regions that cannot be tiled. Thus we must use L_2 if we are to tile X_i nonrigidly. The $1 \times (n-1)$ region directly above the L_2 tile must then be covered with an L_2 tile (see Figure 14).

Figure 14

Tiling the n-Square Xᵢ in Q₁ by T⁺(C₂, n, n), n ≥ 3

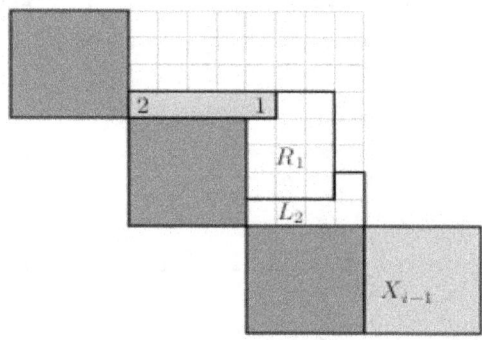

Consider now what tile covers the lower left corner of X_{i+1}, cell 2 in Figure 14. A brief analysis which we will skip shows that any possible covering of cell 2 by L_1, L_2, R_1, R_2, results in the creation of a region of width 1 between the cells 1 and 2 that cannot be completely covered. If cell 2 is tiled by S, then we have a contradiction similar to that in Figure 13, b). So X_i is rigidly tiled, and $T^+(C_2, n, n), n \geq 4$ is rigid with respect to Q_1.

55

The case of Q_3. Consider the base case $k = 1$. Note that we cannot tile the upper right corner with L_1 or L_2, as we would leave regions adjacent to the axes that cannot be tiled. So we attempt to use R_1 or R_2. The same reasoning works for both as they are reflections over the first bisector. We are left with the notch, that has to be covered by L_1, respectively L_2, which leads to a regular tiling of the corner n-square.

Figure 15

Tiling the Corner n-Square in Q_3 by $T^+(C_2, n, n)$, $n \geq 3$

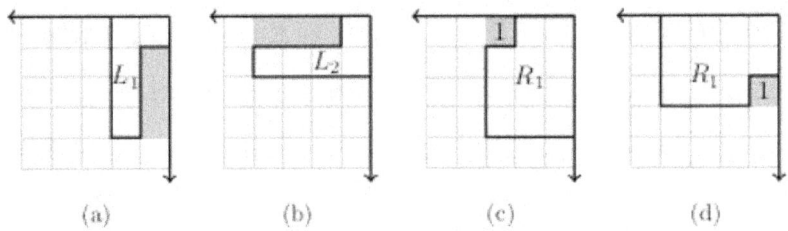

(a) (b) (c) (d)

Now consider $k > 1$ and look at the staircase line in Q_3 (Figure 16). Consider the furthest left not regularly covered n-square adjacent to the staircase, labeled X_i. Its top right corner cell cannot be covered by L_1 or L_2 by the same reasons as shown in Figures 15, a), b). For the same reason as shown in Figures 15, c), d), if we cover it with R_1 or R_2, it must be regularly covered. So $T^+(C_2, n, n)$, $n \geq 4$, must be rigid with respect to Q_3.

Figure 16

Tiling the n-Square X_i, by $T^+(C_2, n, n)$, $n \geq 3$

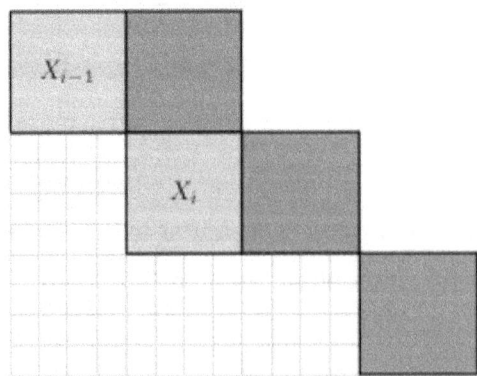

4. Tiling Q_1 by $T^+(C_1, mn, n)$

In this section we prove Theorem 1, 1). We consider the tiling of Q_1 by $T^+(C_1, mn, n)$, $m \geq 2, n \geq 3$. In the diagrams $n = 3, m = 2$, but the arguments are general.

Definition 2. An L_1 tile that is part of a tiling of Q_1 by $T^+(C_1, mn, n)$ is said to be in an irregular position if the coordinates of its lowest left corner are both divisible by n, and if all n-squares below and to the left of its lowest left corner are regularly covered. The corresponding notion for an L_2 tile is defined via a reflection about the line y = x.

The L_1 tile in Figure 18, a), is in an irregular position. All dark gray n-squares are regularly covered.

Definition 3. Assume that Q_1 is tiled by $T^+(C_1, mn, n)$. A gap is a rectangular region of height n in the square lattice that satisfies the following properties:

1. The coordinates of the lower left corner are divisible by n.
2. The n-squares on the left side and below the upper level of the gap are regularly covered.
3. The n-squares directly below the gap (which in the case of gaps of length not divisible by n include a rightmost n-square that has less than n cells directly below the gap) are regularly covered.
4. The lower length 1 part of the right side of a gap of length not divisible by n is not covered by tiles from the tiling of Q_1.
5. The lower length n − 1 part of the right side of a gap of length divisible by n is not covered by tiles from the tiling of Q_1.

Pictures of gaps are shown in Figure 17. The dark gray n-squares are regularly covered. Note the thick segments on the right sides of the gaps of length 1, that cannot be covered by tiles from the tiling of Q_1. We do not assume anything about the tiling of the remaining white regions in Q_1 or of the gaps.

Figure 17

Pictures of Gaps for Q_1

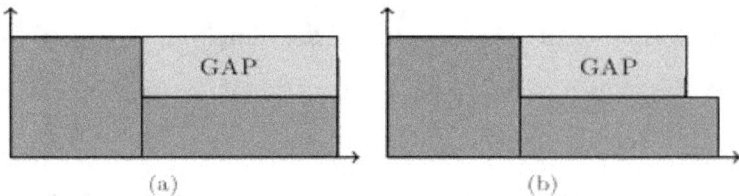

(a)

(b)

Figure 18

An L_1 Tile in an Irregular Position and the Staircase Line for Q_1

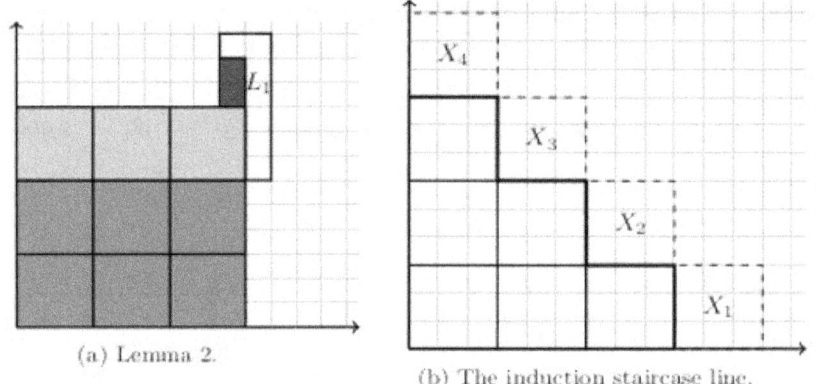

(a) Lemma 2.

(b) The induction staircase line.

Figure 19

An R₁ Tile Covers the Lower Left Cell in the Gap – the Base Case

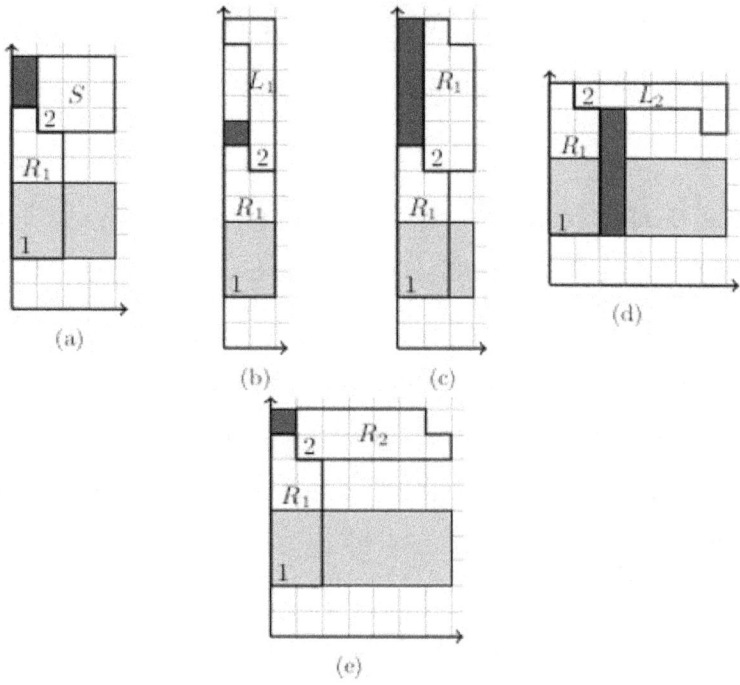

For the following two lemmas we assume that a tiling of Q_1 by $T^+(C_1, mn, n)$ is given.

Lemma 1. Assume that there exists a gap of length $L \geq 1$ for which the leftmost n-square is not regularly covered. Then there exists an L_1 tile in an irregular position that is above the bottom of the gap and to the left of the right side of the gap.

Proof. If $L = 1$ we are done, as the gap can be filled only by an L_1 tile in an irregular position. Let d be the distance between the right side of the gap and the y-axis. We proceed by induction on d. For the induction step, we show that an L_1 tile, or a new gap of length $L \geq 1$ with the leftmost n-square not regularly covered, appears that is above or on the left side of the gap. In the diagrams the gaps are colored in light gray and the regions that cannot be covered completely by $T^+(C_1, mn, n)$ are colored in dark gray.

We consider first the case when the left side of the gap is based on the y-axis. Here we will finish with a contradiction. This includes the base case $d = 2$ for the induction. Look at the tiling of cell 1, the lowest leftmost in the gap. We cannot use an S tile because of the hypothesis. We cannot use a L_1, L_2 tile because the gap is too close to the y-axis. So we can use only a R_1 or a R_2 tile to cover cell 1.

If we use a R_1 tile, then the leftmost n-square in the gap is regularly covered, in contradiction to our assumption. See Figure 19 for the diagrams of the cases that appear when we try to cover cell 2, the notch of the R_1 tile. If $n > 3$, in cases a), b), c), e) only vertical L_1 tiles can be introduced in the dark gray regions and they leave certain cells uncovered. In case d), due to the fact that the right most n-square below the gap is regularly covered, only a L_1 tile may fit to cover the dark gray cell immediately below the L_1 tile. But then the remaining of the dark gray region sits between the R_1 tile and the L_1 tile and cannot be covered.

If cell 1 is covered by an R_2 tile then cell 2, the notch of the R_2 tile, cannot be covered without forcing the leftmost n-square in the gap to follow the rectangular pattern or leading to a contradiction. See Figure 20 for the diagrams. Figures a) through e) show what happens if the length of the gap is not mn. Figure f) shows what happens if the length of the gap is mn. There are fewer cases if the length of the gap is mn due to the higher right edge (of length at least

n − 1) that exists for such gaps. Also, this implies that in Figure f) cell 2 can be covered only by an L_1 tile. In both Figures d) and f) there is no way to cover the dark gray cell adjacent to the L_1 tile.

Figure 20

An R$_2$ Tile Covers the Lower Left Cell in the Gap – the Base Case

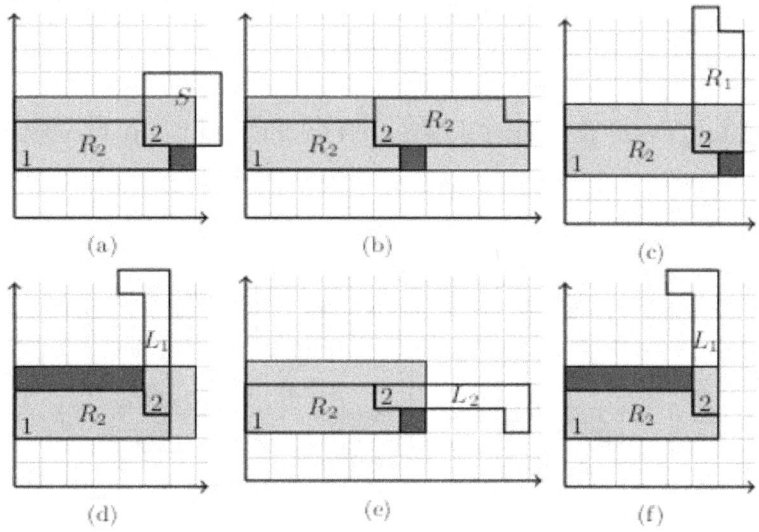

Now, consider the general case. We look at the tiling of cell 1, the lower leftmost in the gap. We cannot use an S tile due to the hypothesis. If we use a L_1 tile, the L_1 tile is in an irregular position. A L_2 tile does not work due to the existence of the right edge of the gap. If we use a R_1 tile, then the n-square containing cell 1 is regularly covered. The reasoning is similar to that done in the case when the left side of the gap is supported by the y-axis, but now there are new gaps with the leftmost n-square not regularly covered appearing. See Figure 21. The case of Figure d) in Figure 20 is identical with the base case.

We observe that the gaps appear either at the level of the R_1 tile, in which case their length can be taken to be divisible by n and the right edge is not covered by any tile, or, if the former is not the case, they appear in the first width n row above the R_1 tile, in which case their length is not divisible by n and the lower length 1 part of the right vertical side of the gap is not covered by tiles.

If we use an R_2 tile, then the n-square containing cell 1 is regularly covered. The reasoning is similar to that done in the case when the left side of the gap is supported by the y-axis and the same dark gray regions as in Figure 20 are impossible to tile.

Figure 21

An R₁ Tile Covers the Lower Left Cell in the Gap – the General Case

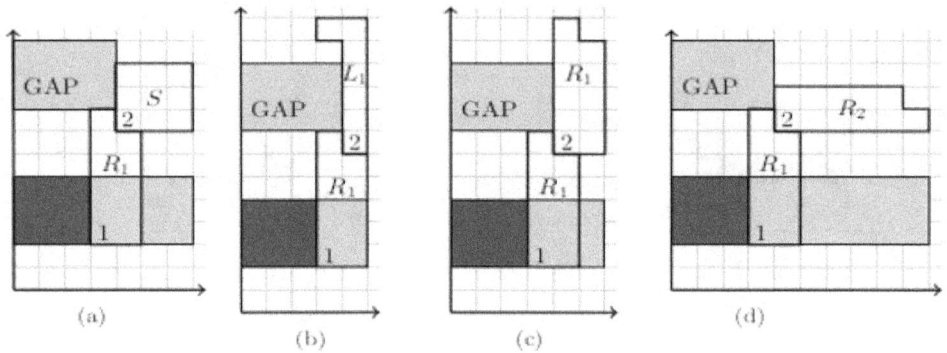

Lemma 2. A tiling of Q_1 by $T^+(C_1, mn, n)$ cannot contain an L_1 or L_2 tile in an irregular position.

Proof. Assume that an L_1 tile is in an irregular position and at minimal distance from the y-axis. Figure 18, a), illustrates the proof of the lemma. The dark gray n-squares are already regularly covered. If not all light gray n-squares are regularly covered, we identify the bottom row of width n in the light gray region in which such an n-square appears and then apply Lemma 1 to a gap in that row. This gives a new L_1 tile closer to the y-axis and leads to a contradiction. If all light gray squares are regularly covered, we observed that the dark gray squares cannot be covered without leading to a contradiction. Indeed, only a horizontal tile L_2 or R_2 may fit in the dark gray region. In either case, if we cover the top dark gray region, there is room left below that tile that cannot be covered. This gives a contradiction. The statement about the L_2 tile follows due to the symmetry of $T^+(C_1, mn, n)$ about the line $y = x$.

62

Proof of Theorem 1. We show that every n-square is regularly covered. We do this by induction on a diagonal staircase of step n, shown in Figure 18, b). We first tile the corner cell of Q_1. The possible cases are similar to those shown in Figures 19, 20 and lead either to a regular tiling of the corner n-square or to a contradiction.

For the induction step we prove that the n-squares X_i in Figure 18, b), immediately above the staircase, are regularly covered. Choose the rightmost square X_i, say X, which is not regularly covered. Note that X is bounded below and n cells to the right by the x-axis or by two n-squares that are regularly covered , and it is bounded to the left by the y-axis or an n-square that is regularly covered. By assumption, bottom left cell, labeled 1, cannot be covered by a S tile. Also, cell 1 cannot be covered by a L_1 or an L_2 tile due to Lemma 2, as the L_1, L_2 tile will be in an irregular position. So cell 1 can be covered only by R_1 or R_2.

If an R_1 tile covers cell 1, then we are either in the case from Figure 19, d), which leads to an immediate contradiction, or in one of the cases that appear in Figure 21. The later cases lead to the appearance of a gap. Now Lemma 1 shows that we have a L_1 tile in an irregular position, which gives a contradiction due to Lemma 2. Note that we did not use in the proof the characterization of X_i as the rightmost n-square that is not regularly covered.

If an R_2 tile covers cell 1, then we observe that the proof done when an R_1 tile covers cell 1 can be repeated verbatim for this case due to the symmetry about the line $y = x$. The presence in this case of extra n-squares regularly covered does not affect the proof.

5. Rigid tiling by $T^+(C_2, mn, n)$

In this section we prove Theorem 1, 5). We consider tilings of Q_2 by $T^+(C_2, mn, n)$, $1 \le i \le 2, m \ge 2, n \ge 3$. The diagrams are drawn for n = 3, m = 2, but the arguments are general.

Definition 4. A L_1 tile that is part of a tiling of Q_2 by $T^+(C_2, mn, n)$ is said to be in an irregular position if the coordinates of its lowest right corner are both divisible by n, and if all n-squares below and to the right of its lowest right corner are regularly covered. The corresponding notion for a L_2 tile is defined via a reflection about the line y = x.

The L_1 tile in Figure 22, a), is in an irregular position. All gray squares are regularly covered.

Figure 22

An L_1 Tile in an Irregular Position and the Staircase Line for Q_2

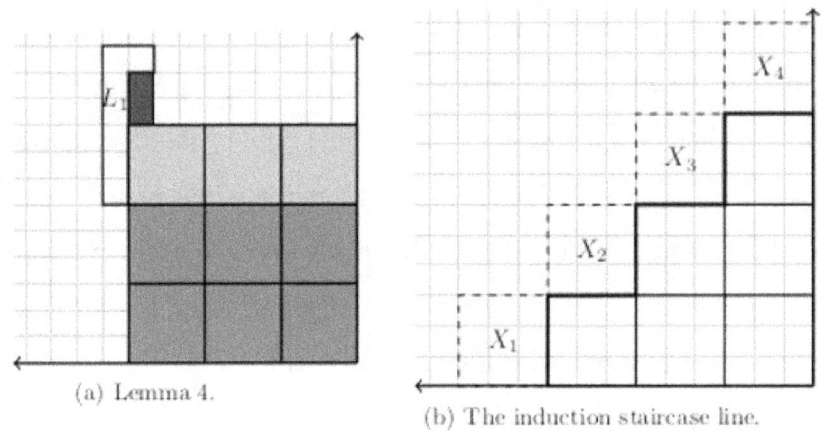

(a) Lemma 4.

(b) The induction staircase line.

Figure 23

Pictures of Gaps for Q_2

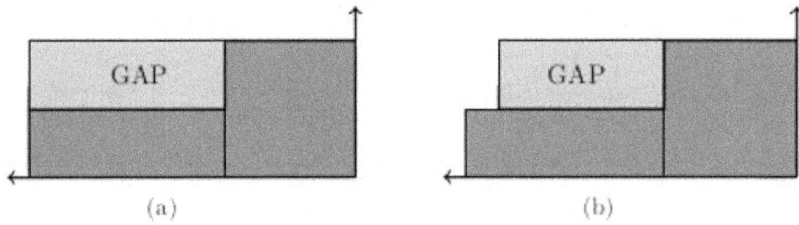

(a)   (b)

Definition 5. Assume that Q_2 is tiled by $T^+(C_2, mn, n)$. A gap is a rectangular region of height n in the square lattice that satisfies the following properties:

1. The coordinates of the lower right corner are divisible by n.
2. The n-squares on the right side and below the upper level of the gap are regularly covered.
3. The n-squares directly below the gap (which in the case of gaps of length not divisible by n include a leftmost n-square that has less than n cells directly below the gap) are regularly covered.
4. The lower length 1 part of the left side of a gap is not covered by tiles from the tiling of Q_2.

64

Pictures of gaps are shown in Figure 23. The dark gray regions follow the rectangular pattern. Note the thick segments on the left sides of the gaps. These cannot be covered by tiles from the tiling of Q_2. We do not assume anything about the tiling of the remaining white regions in Q_2 or of the gaps.

For the following three lemmas we assume that a tiling of Q_2 by $T^+(C_2, mn, n)$ is given.

Lemma 3. Assume that there exists a gap of length L $\geq$ 1 for which the rightmost n-square is not regularly covered. Then there exists an L_1 tile in an irregular position that is above the bottom of the gap and to the right of the left side of the gap.

Proof. If $L = 1$ we are done, as the gap can be filled only by a L_1 tile in an irregular position. Let d be the distance between the left side of the gap and the y-axis. We proceed by induction on d. For the induction step, we show that a L_1 tile, or a new gap of length $L \geq 1$ with the rightmost n-square not regularly covered, appears that is above or on the right side of the gap. In the diagrams the gaps are colored in light gray and the regions that cannot be covered completely by $T^+(C_2, mn, n)$ are colored in dark gray.

We consider first the case when the right side of the gap is based on the y-axis. Here we will finish with a contradiction. This includes the base case $d = 2$ for the induction. Look at the tiling of cell 1, the lowest rightmost in the gap. We clearly cannot use a R_2 tile. We cannot use a S tile because of the hypothesis. We cannot use an L_1 tile because the gap is too close to the y-axis. If we use a R_1 tile, then the rightmost n-square in the gap is regularly covered, in contradiction to our assumption. See Figure 24 for the diagrams of the cases that appear when we try to cover cell 2, the notch of the R_1 tile. In Figures a), b), c) in the gray regions we can fit only L_1 tiles. This leaves some room among the tiles and the y-axis. In Figures d), e), f) the gray regions can be covered only by horizontal tiles L_2, R_2. The tiles have to be paired and tile $n \times k$ rectangles, as otherwise there will be regions between the tiles that cannot be covered. As neither one of the dark gray regions has height divisible by n, we conclude that covering them is impossible.

Figure 24

A R_1 Tile Covers the Lower Left Cell in the Gap – the Base Case

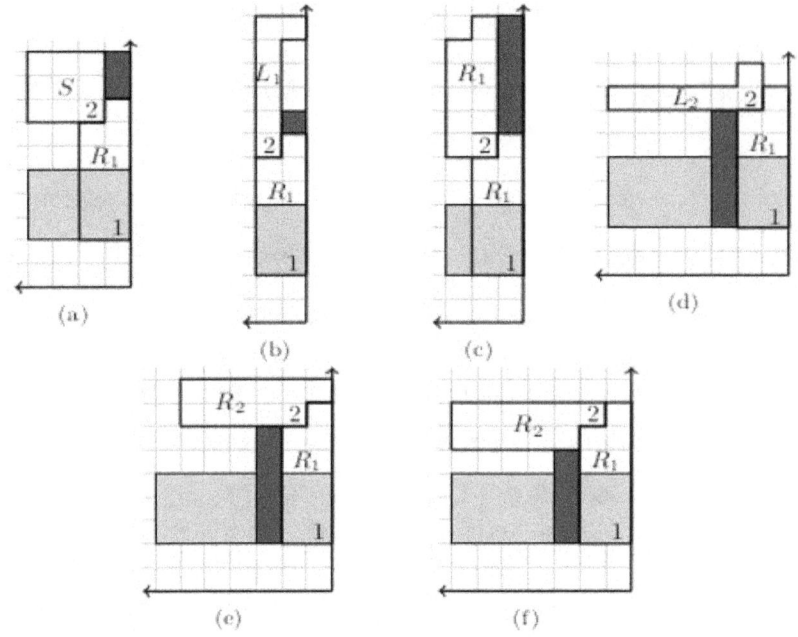

Figure 25

An L_2 Tile Covers the Lower Left Cell in the Gap – the Base Case

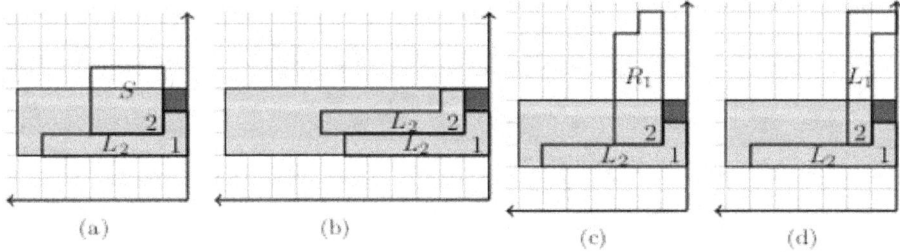

Figure 26

A R_1 Tile Covers the Lower Left Cell in the Gap – the General Case

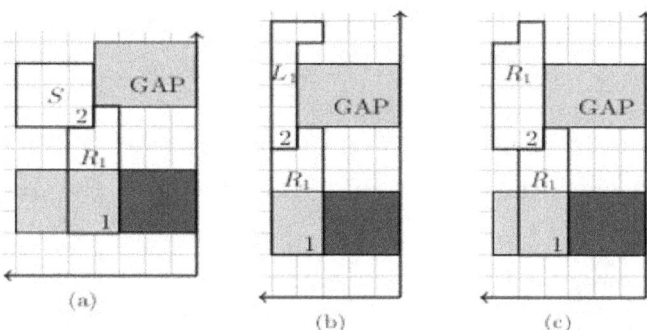

If cell 1 is covered by an L_2 tile then cell 2, diagonally adjacent to cell 1 cannot be covered without forcing the rightmost n-square in the gap to be regularly covered or leading to a contradiction. See Figure 25 for the diagrams. In all cases in the gray regions we can fit only L_1 tiles. This leaves some room among the tiles and the y-axis.

Consider now the general case. We look at the tiling of cell 1, the lower rightmost in the gap. We cannot use an S tile due to the hypothesis. If we use an L_1 tile, it will be in an irregular position. A R_2 tile clearly does not work. If we use a R_1 tile, then the n-square containing cell 1 is regularly covered. The reasoning is similar to that done in the case when the left side of the gap is supported by the y-axis. The possible ways to cover cell 2, the notch of the R_1 tile, are similar to those shown in Figure 24.

In the cases d), e) and f) similar gray regions cannot be tiled. In the cases a), b), c) there are new gaps appeared as shown in Figure 26. We observe that the gaps appear either at the level of the R_1 tile, in which case their length can be taken to be divisible by n and the right edge is not covered by any tile, or they appear in the row of width n above the R_1 tile, in which case their length is not divisible by n and the lower length 1 part of the right vertical side of the gap is not covered by tiles. If we use an L_2 tile, then the n-square containing cell 1 is regularly covered.

The reasoning is similar to that done in the case when the right side of the gap is supported by the y-axis. We look at the dark squares (regions) in Figure 25. In case d), even if the gap is away from the y-axis, there no way to cover the dark gray square. In cases a), b), c), the dark gray square can be covered only by the long end of a L_1 tile, see Figure 27, a).

Figure 27

An L_2 Tile Covers the Lower Left Cell in the Gap – the General Case

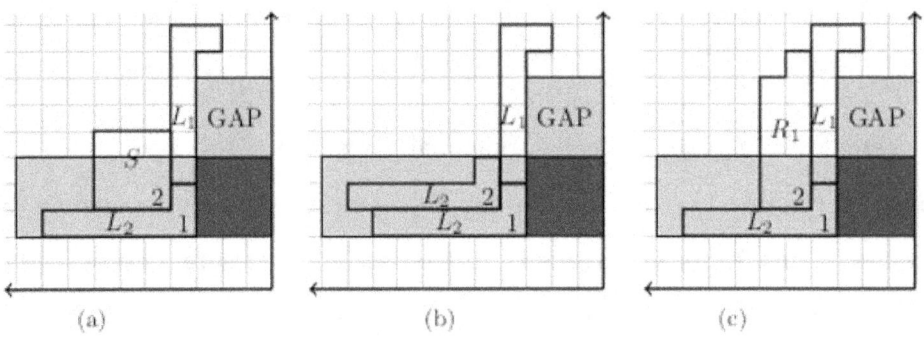

(a) (b) (c)

We check then for the appearance of a gap on the right side of the L_1 tile that does not have the right most n-square regularly covered. If such a gap appears, we are done. If not, all n-squares between the y-axis and the L_1 tile are regularly covered and we need to use a vertical L_2 tile to cover the cell under the short end of the L_1 tile, see Figure 28, a).

This creates a gap of length divisible by n on the right side of the L_2 tile, see Figure 28, a). If this new gap has the rightmost n-square not regularly tiled, we are done. If not, we observe that any covering of cell 1 leads either to an immediate contradiction or forces cell 2 to be covered by a L_1 tile. A pattern appears and continuing in this manner leads to the appearance of a gap that has the right most n-square not tiled regularly or leads to the appearance of a new staircase build out of L_1, L_2 tiles, see Figure 28, b). The staircase eventually reaches the -axis leading to a contradiction.

Figure 28

An L_2 Tile Covers the Lower Left Cell in the Gap – The General Case-Continuation

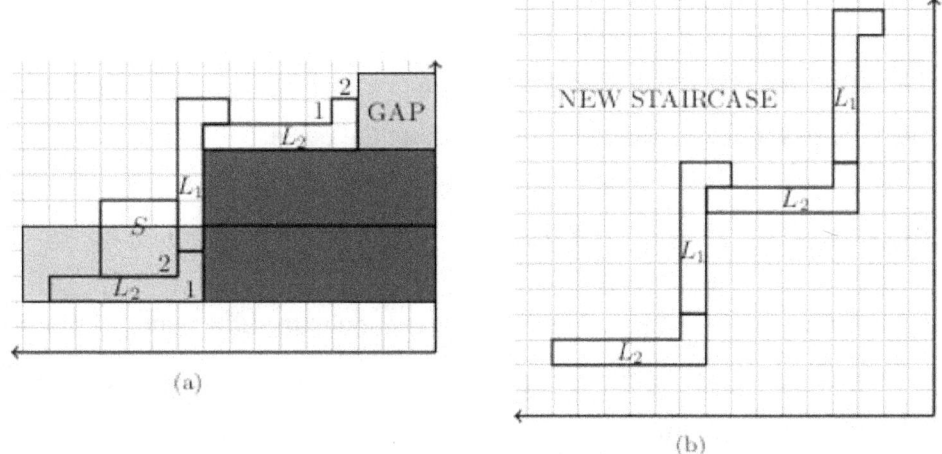

(a)   (b)

Lemma 4. A tiling of Q_2 by $T^+(C_2, mn, n)$ cannot contain an L_1 or L_2 tile in an irregular position.

The proof of Lemma 4 is similar to that of Lemma 2 and left for the reader.

Proof of Theorem 1. We show that every n-square follows the rectangular pattern. We do this by induction on a diagonal staircase of step n, shown in Figure 22, b). We first investigate the tiling of the corner cell of Q_1. The possible cases are similar to those shown in Figures 24, 25 and lead either to a regular tiling of the corner n-square or to a contradiction.

For the induction step we prove that the n-squares X_i in Figure 22, b), follow the rectangular pattern. Choose the rightmost square X_i, say X, which does not follow the rectangular pattern (see Figure 22, b)). Note that X is bounded below and n cells to the right by the x-axis or by two n-squares that are regularly covered, and it is bounded to the left by the n-axis or an n-square that is regularly covered. By assumption, bottom left cell, labeled 1, cannot be covered by an $|S$ tile. Also, cell 1 cannot be covered by $L_1(L_2)$ due to Lemma 4, as the $L_1(L_2)$ tile is in an irregular position. So cell 1 can be covered only by R_1 or R_2.

If a R_1 tile covers cell 1, then we are either in the cases that appear in Figures 24, d), e), f), or in those in Figure 26. The later cases lead to the appearance of a gap. Now Lemma 3 shows that we have a L_2 tile in an irregular position, which gives a contradiction due to Lemma 4.

If a R_2 tile covers cell 1, then we observe that the proof done when a R_1 tile covers cell 1 can be repeated verbatim for this case due to the symmetry about the line $y = x$. The presence in this case of extra n-squares regularly covered does not affect the proof.

6. Nonrigid results for $T^+(C_1, mn, n)$.

In order to prove the nonrigid results in Theorem 1 it is enough to show nonrigid tilings of Q_2, Q_3 by $T(C_1, mn, n)$. We use repeatedly in the proofs that a multiple of a rectangle of size $mn \times n$ or $n \times mn$, or an half-infinite strip of width a multiple of n can be tiled by the tile set.

In Figure 29, a), we show a nonrigid tiling of the second quadrant by $T(C_1, mn, n)$. Regions I and II are infinite strips of width mn and region III is a copy of the second quadrant.

In Figure 29, b) we show a nonrigid tiling of the third quadrant by $T(C_1, mn, n)$. Regions I and II are infinite strips of width mn, region III is a copy of the second quadrant, region IV is a rectangle of size $(m - 1)n \times mn$ and region V is a rectangle of size $mn \times (m - 1)n$.

Figure 29

Non-Rigid Tilings by T(C₁, mn, n)

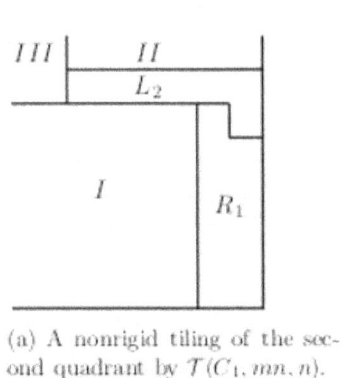

(a) A nonrigid tiling of the second quadrant by $T(C_1, mn, n)$.

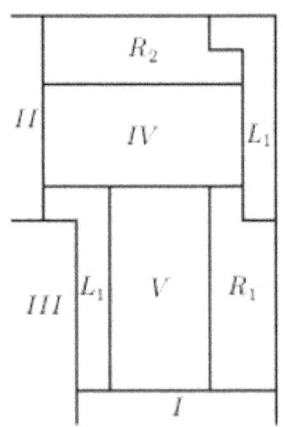

(b) A nonrigid tiling of the third quadrant by $T(C_1, mn, n)$.

7. Nonrigid tilings for T(C_2, 3, 3).

In Figure 30, a), we show a nonrigid tiling of the second quadrant by T(C_2, 3, 3). Regions I, III and IV are half infinite strips of width 3 and region II is a copy of the second quadrant.

In figure 30, b), we show a nonrigid tiling of the third quadrant by T(C_2, 3, 3). Regions II and IV are half infinite strips of width 3, region I is a half infinite strip of width 6 and region III is a copy of the second quadrant.

Figure 30

Non-rigid Tilings by T(C_2, 3, 3)

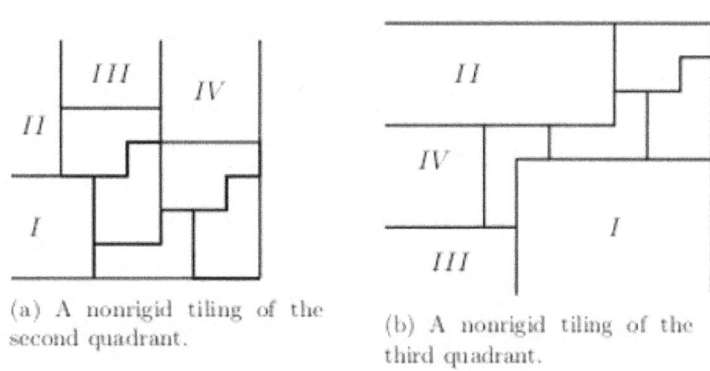

(a) A nonrigid tiling of the second quadrant.

(b) A nonrigid tiling of the third quadrant.

8. Nonrigid results for coprime dimensions.

In this section we prove Theorem 3. Due to the symmetries present in the tiling sets, it is enough to show the following:

1. Nonrigid tilings for $T(C_1, k, n)$ for the first, second and third quadrants;

2. Nonrigid tilings for $T(C_2, k, n)$ for the first, second and third quadrants.

We first give a proof for 2. and later a proof of 1. We will use that a multiple of a $k \times n$ or $n \times k$ rectangle, or a half infinite strip of width k or n can be tiled by $T(C_i, k, n)$, $1 \le i \le 2$. \$ Recall that if k, n are coprime, there exists positive integers x, y such that $yk - xn = 1$.

Proof of 2. Figure 31, a), shows a nonrigid tiling of the first quadrant by $T(C_2, k, n)$. For x, y positive integers such that $xk - yn = 1$, region I is an infinite strip of width xk, region II is an infinite strip of width k, region III is an infinite strip of width $(k-1)yn$, region IV is a copy of Q_1 and region V is a rectangle of size $(k-1)kx \times yn$.

Figure 31

Non-rigid Tilings by T(C_1, k, n) and T(C_2, k,n) for n,k Coprime

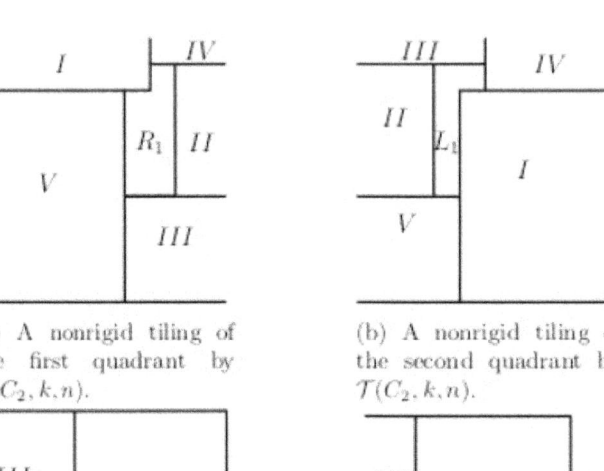

(a) A nonrigid tiling of the first quadrant by $T(C_2, k, n)$.

(b) A nonrigid tiling of the second quadrant by $T(C_2, k, n)$.

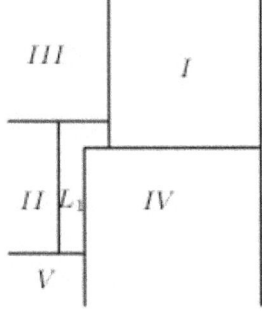

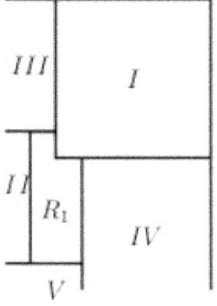

(c) A nonrigid tiling of the third quadrant by $T(C_2, k, n)$.

(d) A nonrigid tiling of the third quadrant by $T(C_1, k, n)$.

72

Figure 31, b), shows a nonrigid tiling of Q_2 by $T(C_2, k, n)$. Let x, y, z, w positive integers such that $xn - yk = 1, zk - wn = 1$. Region I is a rectangle $zk(n - 1) \times xn$, region II is an infinite strip of width k, region III is a copy of Q_2, region IV is an infinite strip of width yk, and region V is an infinite strip of width $wn(n-1)$.

Figure 31, c), shows a nonrigid tiling of the third quadrant by $T(C_2, k, n)$. For x, y positive integers such that $xk - yn = 1$, region I is a rectangle of size $xk \times yn$, region II is an infinite strip of width k, region III is an infinite strip of width yn, region IV is an infinite strip of width xk and region V is a copy of Q_3.

Observe that the tile L_2 in $T(C_1, k, n)$ can be obtained via a counterclockwise rotation by 90^0 from L_1 in $T(C_2, k, n)$. Using 2, this gives nonrigid tilings by $T(C_1, k, n)$ for Q_1, Q_2. Figure 31, d), shows a nonrigid tiling of Q_3 by $T(C_1, k, n)$.

Let x, y, z, w be positive integers such that $xn - yk = 1, zk - wn = 1$. Region I is a rectangle $xn \times (n - 1)kz$, region II is an infinite strip of width k, region III is an infinite strip of width yk, region IV is an infinite strip of width $(n-1)mw$, and region V is a copy of Q_3.

9. Signed tilings

In this section we prove Theorem 2. We recall that signed tilings are finite placements of tiles on a plane, with weights +1 or -1 assigned to each of the tiles. We say that they tile a region R if the sum of the weights of the tiles is 1 for every cell inside R and 0 for every cell elsewhere. In order to prove Theorem 2 it is enough to show a signed tiling of a single cell by our tiling set.

We show the geometric proof only for the tile set $T(C_1, 5, 5)$ in Figure 32. Proofs in the other cases are immediate modifications of this one. We start by placing a notched rectangle on the grid, then use the L-shaped complement to construct a bar of size n, and then use the bar and the symmetric about the first bisector of the L-shaped complement to obtain the single cell.

Figure 32

A Signed Tiling of a Cell

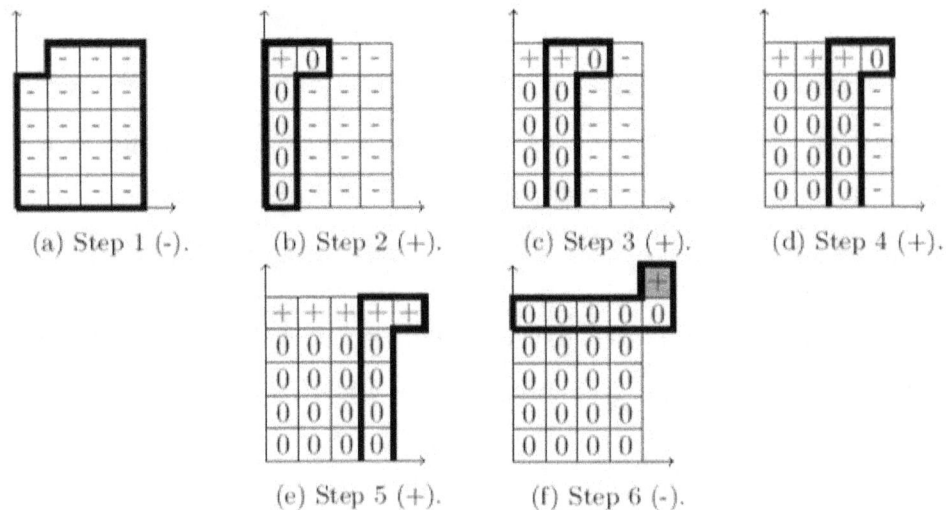

(a) Step 1 (-).   (b) Step 2 (+).   (c) Step 3 (+).   (d) Step 4 (+).

(e) Step 5 (+).   (f) Step 6 (-).

Acknowledgments. This paper was written during the Summer 2013 program REU at Pennsylvania State University, supported by the NSF grant DMS-0943603. A. Calderon, S. Fairchild, and S. Simon are undergraduate students. V. Nitica was one of the faculty coordinators. He was partially supported by Simons Foundation Grant 208729.

74

Bibliography

[1] N.G. de Brujin, Filling boxes with bricks, Amer. Math. Monthly, **76** (1969), 37--40.

[2] M. Chao, D. Levenstein, V. Nitica, R. Sharp, A coloring invariant for ribbon L-tetrominoes, Discrete Mathematics, **313** (2013) 611--621.

[4] R. Hochberg, The gap number of the T-tetromino, Discrete Math., **338** (2015), 130--138.

[5] S.W. Golomb, Checker boards and polyominoes, Amer. Math. Monthly, **61** (1954) 675--682.

[6] S.W. Golomb, Replicating figures in the plane, Mathematical Gazette, **48** (1964) 403--412.

[7] S.W. Golomb, Tiling with polyominoes, Journal of Comb. Theory, **1** (1966), 280--296.

[8] S.W. Golomb, Tiling with sets of polyominoes, Journal of Comb. Theory, **9** (1970), 60--71.

[9] S.W. Golomb, Polyominoes, Puzzeles, Patterns, Problems, and Packings (2-nd ed.), Princeton University Press, NJ, 1994.

[10] V. Nitica, Rep-tiles revisited, in the volume MASS Selecta: Teaching and Learning Advanced Undergraduate Mathematics, American Mathematical Society, 2003.

[11] V. Nitica, Any tiling of the first quadrant by ribbon L-shaped n-ominoes follows the rectangular pattern, Open Journal of Discrete Mathematics, **5** (2015) 11--25 .

[12] I. Pak, Ribbon tile invariants, Trans. Amer. Math. Society, **352** (2000) 5525--5561.

[13] M. Reid, Many L-shaped polyominoes have odd rectangular packings, Annals of Combinatorics, **18** (2014) 341--357.

[14] L. Sallows, On self-tiling tile sets, Mathematics Magazine, **85** (2012) 323--333.

[14] L. Sallows, More on self-tiling tile sets, Mathematics Magazine, **87** (2014) 10

BOOK REVIEWS

Edited by:Charles Ashbacher

Charles Ashbacher Technologies

5530 Kacena Ave

Marion, IA 52302

E-mail: cashbacher@yahoo.com

The Early Mathematics of Leonhard Euler, edited by C. Edward Sandifer, The Mathematical Association of America, Washington, D. C. 2007. 393 pp., $25.95(hardbound). ISBN 0883855593.

Leonhard Euler was certainly one of the most prolific mathematicians of all time, as Sandifer states in the preface, his exact rank depends on how you define the nature of the publications. There is a brief discussion of the ranking of Erdös versus Euler based on how coauthors are scored. What is not a matter of dispute is who was the dominant mathematician of the middle of the eighteenth century. In that time, Euler had no real peer.

This book is a listing and analysis of the first mathematical papers that Euler had published, they are sequentially segmented by year or a short interval of years. The first timeframe is 1725-1727 and the last is 1741. Approximately 50 papers are examined, each is summarized and annotated with explanations of the content and the historical value it had. A few considered to be the major ones of historical significance are given asterisks for emphasis.

Since these are all his early papers, some are fairly nondescript and even Sandifer says so. What is clear from these papers is a clear development of mathematics, both in content and in the notational representation. One paper contains the first use of now standard $f(x)$ notation for functions and others the development of new techniques used in proofs. When he was wrong, and even Euler was wrong, he advanced the field of mathematics. When you embark on even a rudimentary examination of mathematical history you realize how valuable modern notation is compared to the relatively crude forms used several hundred years ago.

Euler was truly a giant in mathematics, in this book you can read about the early emergence of his work. It sounds odd to say it but these fifty papers are just the preliminary material of his output. As singer Carly Simon put it, "Nobody does it better."

Charles Ashbacher

Mathematical Sorcery, by Calvin C. Clawson, Perseus Publishing, Cambridge, MA, 1999. 294 pp., $16.50 (paper) ISBN 073820496X.

This book is a popular history of mathematics with a catchy and slightly disingenuous title. For there is no sorcery or magic involved, it is the wonder and usefulness of mathematics in both the pure and applied form.

It begins with the emergence of the concept of counting and the historical context as to why it emerged. With the development of agriculture, both animal husbandry and crops, human groups became fixed in location. This led to the emergence of governments to organize and protect the populations, which led to the levying of taxes. All of this required the ability to count, tally and record, which meant the mental concept of numbers had to develop as well as a way to efficiently record and manipulate them. The last chapter of history covers the development of and basic applications of calculus, so the history essentially ends in the first half of the eighteenth century.

The journey from start to finish is an understandable tour through several of the most significant advances in mathematics, from the development and use of negative numbers, fractions, irrational and transcendental numbers to complex numbers. Clawson is to be commended for he does not skimp on the use of formulas, when one is needed one is used.

Math, science and the increasing complexity of societies have been married into a feedback loop for thousands of years. Sometimes the need led to the development of mathematics, for example, when society needed counting numbers, the math was invented. Other times the math had to be invented to explain the science. In other circumstances, the math was developed before society had a use for it.

In all cases, the development of the math proceeded and Clawson does an excellent job in explaining the new math concepts, the reason it was developed and the niche it filled in society.

Charles Ashbacher

Curves and Approximations, by John R. Hendricks, privately published by the author, 1999. 36 pp., (stapled). ISBN 096847005x.

This short book contains some very interesting curves that are generally extensions of well known ones. The formulas used to create the curves are included as well as some BASIC code to plot them.

The cosine nodosus is a cosine curve that has a loop bisected by the y-axis in the middle and the binodal lemniscate is like a lemniscate but has an ovoid in the middle with a node on each end.

The trinodal lemniscate has two loops in the middle with nodes on each end and the quadnodal lemniscate has three loops in the center.

Unusual curves have the additional property that they are interesting to look at. John R. Hendricks did a great deal of work in magic squares, cubes and tesseracts. In this book he adds some unusual and attractive curves to his extensive list of accomplishments.

Charles Ashbacher

Magic Squares to Tesseracts by Computer, by John R. Hendricks, privately published, Victoria, B. C., Canada, 1998. ISBN 0968470009.

The first section of the book is a restatement of well known information about magic and Latin squares as well as modular arithmetic and the basics of creating fundamental magic squares. The topic of chapter 4 is "Magic Squares of Order 4k" and the rules that are used to create them.

The real excitement begins in chapter 5, where Hendricks develops the three-dimensional magic cube. Rules for their development and several examples of such squares are given, including a perfect magic cube of order 9 with magic sum 3285.

Chapter 10 introduces the 4th-dimension and the tesseract, the extension of the cube into four dimensions. One of the main points of difficulty is developing a notation for representing such structures. This representation is both visual as well as formulaic. I will make no attempt to reproduce the images of the tesseracts. Hendricks uses a set of ordered 4-tuples with the numbers 1 through 3 to represent the nodes of the tesseracts. Given that the simplest such tesseract has three numbers along each direction, there are $3 * 3 * 3 * 3 = 81$ numbers in the smallest one, specifically the numbers 1 through 81. An example of an order 3 magic tesseract appears on page 112. Additional examples of other tesseracts are also given.

John R. Hendricks was one of the major figures in extending the mathematics of magic squares far beyond the widely known basics. His work involved a combination of imagination and persistence, two characteristics easily seen in this book.

Charles Ashbacher

Note: John R. Hendricks passed away in 2007, so the previous two books may no longer be available.

Quantitative Methods: An Introduction for Business Management 1st Edition,by Paolo Brandimarte, John Wiley & Sons, Hoboken, New Jersey, 2011. 912 pp., $140.00 (hardbound). ISBN 978-0-470-49634-3.

This is a math book with some business and economic examples included. A better subtitle and explanation would be "The math you need to know to use and understand advanced quantitative methods in business." The book is split into sections and a look at the 16 chapter titles after the introductory first chapter will explain a great deal. They are

*) Calculus

*) Linear algebra

*) Descriptive statistics

*) Probability theories

*) Discrete random variables

*) Continuous random variables

*) Dependence, correlation and conditional expectation

*) Inferential statistics

*) Simple linear regression

*) Deterministic decision models

*) Decision making under risk

*) Multiple decision makers, subjective probability and other wild beasts

*) Introduction to multivariate analysis

*) Advanced regression models

*) Dealing with complexity: Data reduction and clustering

The author presupposes that the reader is very familiar with the topics before the chapter is encountered. There is no slack cut in terms of the level of difficulty of the mathematics. When a topic is first stated, it is as a reminder and not as a review. As the list and number of pages indicate, this is a book for a two or three semester course sequence in the mathematics used in high level quantitative methods used in business.

Fundamentally, it is a text for courses designed for people that are math majors moving into quantitative business management. For that, it is an excellent book. Unlike some other books, no one can criticize this one for being weak on the math, I would suspect that the students using it would argue the contrary.

Charles Ashbacher

Statistical Models: Theory and Practice, by David A. Freedman, Cambridge University Press, New York, New York, 2005. 424 pp., $54.99 (paper). ISBN 978-0521671057.

While in many ways this is a book of the mathematics used in the construction of statistical models, there are some gems at the end. The first chapter is very educational as it contains explanations of three of the best experiments ever conducted, some of which were natural. A natural experiment is where data is collected and then assigned to treatment or control in a random manner. The data is then analyzed and then processed in order to better understand or to assign an explanatory mathematical model.

The first is the Health Insurance Plan (HIP) study regarding the efficacy of breast cancer treatments. The second is the famous data analysis of the spread of cholera conducted by John Snow in 1855, decades before the emergence of the germ theory of disease. The last is a description of the model of poverty developed by G. U. Yule in the last year of the nineteenth century. Using census data, he developed a model on the causes of poverty. These three examples serve as a primer on how valuable statistical models can be and how they are derived from databases.

The titles of chapters 2 through 8 explain the mathematical contents fairly well. They are:

*) The Regression Line

*) Matrix Algebra

*) Multiple Regression

*) Path Models

*) Maximum Likelihood

*) The Bootstrap

*) Simultaneous Equations

The math is all soundly developed so that the reader will understand how it is used to create the models.

However, I found the reprints in the appendix to be by far the most interesting content. There are four of them and the first is of a paper by James L. Gibson where he examines the sources of political repression during the McCarthy era. Gibson investigates whether the primary source of repression was the political elite or from the mass public.

The second reprint is of a paper by William N. Evans and Robert M. Schwab and is an examination of the relative effectiveness of public and Catholic high schools regarding the students finishing high school and starting college. The third reprint is of a paper by Ronald R. Rindfuss, Larry Bumpass and Craig St. John and is an examination of the relationships between the education that a woman has versus her rates of bearing children. The last reprint is of a paper by Mark Schneider, Paul Teske, Melissa Marschall, Michael Mintrom and Christine Roch. It is an examination of whether the opportunity for parents to select the public schools their students attend leads to their being more involved in school programs such as the PTA.

Reading these papers gives the reader an appreciation for the breadth of use that mathematical and statistical models can be applied to. In a world where people cannot be assigned or manipulated, only the power of statistical modeling can be used to evaluate and explain the consequences of aspects of public policy.

Charles Ashbacher

The Harmony of the World: 75 Years of *Mathematics Magazine*, edited by Gerald L. Alexanderson and Peter Ross, the Mathematical Association of America, Washington, D. C., 2007. 304 pp., $55.95 (hardbound). ISBN 9780883855607.

Given a publication history of 75 years, finding a collection of 38 papers that are an accurate representation of the content is an impossible task. Mathematics is a very broad subject with many applications that have been reflected in the pages of **Mathematics Magazine**. Yet, to their credit, the editors do as good a job as is possible.

The papers in this book, all of which appeared in **Mathematics Magazine**, are yet another demonstration of the incredible breadth of mathematics. Although the majority deal with some aspect of history, the study of the history of mathematics is also a demonstration of the breadth.

The organization is based on decades, with a small number of papers from each of the ten-year spans. For example, there are three papers from the 1940s, five papers from the 1950s and seven papers from the 1960s. This is an organizing principle only, there is no other significance that can be attached to it.

Although some of the papers contain a bit of rigor, all of the papers are expository in nature, accessible to the advanced undergraduate and suitable for assigned reading. Teachers of the history of mathematics will find many of the papers appropriate for student reading and discussion. Each paper begins with a short "Editor's note" that establishes the context of the paper, including a brief bio of the author. As with the rest of the book, this feature is also well done.

Charles Ashbacher

Euclidean and Transformational Geometry: A Deductive Inquiry, by Shlomo Libeskind, Jones and Bartlett Publishers, Sudbury, Massachusetts, 2008. 371 pp., $231.95 (hardbound). ISBN 9780763743666.

Despite having a title that is a bit different, this is a textbook in basic Euclidean geometry. The emphasis is on demonstrating the theorems and other concepts via diagrams, formulas are used only when it is necessary to do so. Segments containing different treatments are color-coded, for example the statement and proof of a theorem is placed within a blue backdrop.

The level of discourse is kept at a modest level, the book was developed for the instruction of students in a college geometry course where their goal after graduation is to teach high school mathematics. It clearly is well suited for that course, yet it could also be used as a text in a more advanced high school geometry class. I have tutored a few high school geometry students and much of the content in this book is what we worked on. A set of problems is given at the end of each section and hints and answers to some of them are given at the end.

If you teach high school math or a course designed to train those that will do so in the future, this is a book that will be an effective text from the standpoint of both the teacher and the student.

Charles Ashbacher

SOLUTIONS TO PROBLEMS FROM TRM VOLUME 2

Lamarr Widmer

Messiah College,
Suite 3041, One College Avenue,
Mechanicsburg, PA 17055

1. Mighty Magic by Clarence Gipbsin, Represa, CA

How many magic squares are present in Figure 1, what are their orders and how are they embedded in the 27 by 27 square?

Figure 1

638	71	476	631	64	469	636	69	474	575	8	413	568	1	406	573	6	411	620	53	458	613	46	451	618	51	456
233	395	557	226	388	550	231	393	555	170	332	494	163	325	487	168	330	492	215	377	539	208	370	532	213	375	537
314	719	152	307	712	145	312	717	150	251	656	89	244	649	82	249	654	87	296	701	134	289	694	127	294	699	132
633	66	471	635	68	473	637	70	475	570	3	408	572	5	410	574	7	412	615	48	453	617	50	455	619	52	457
228	390	552	230	392	554	232	394	556	165	327	489	167	329	491	169	331	493	210	372	534	212	374	536	214	376	538
309	714	147	311	716	149	313	718	151	246	651	84	248	653	86	250	655	88	291	696	129	293	698	131	295	700	133
634	67	472	639	72	477	632	65	470	571	4	409	576	9	414	569	2	407	616	49	454	621	54	459	614	47	452
229	391	553	234	396	558	227	389	551	166	328	490	171	333	495	164	326	488	211	373	535	216	378	540	209	371	533
310	715	148	315	720	153	308	713	146	247	652	85	252	657	90	245	650	83	292	697	130	297	702	135	290	695	128
593	26	431	586	19	424	591	24	429	611	44	449	604	37	442	609	42	447	629	62	467	622	55	460	627	60	465
188	350	512	181	343	505	186	348	510	206	368	530	199	361	523	204	366	528	224	386	548	217	379	541	222	384	546
269	674	107	262	667	100	267	672	105	287	692	125	280	685	118	285	690	123	305	710	143	298	703	136	303	708	141
588	21	426	590	23	428	592	25	430	606	39	444	608	41	446	610	43	448	624	57	462	626	59	464	628	61	466
183	345	507	185	347	509	187	349	511	201	363	525	203	365	527	205	367	529	219	381	543	221	383	545	223	385	547
264	669	102	266	671	104	268	673	106	282	687	120	284	689	122	286	691	124	300	705	138	302	707	140	304	709	142
589	22	427	594	27	432	587	20	425	607	40	445	612	45	450	605	38	443	625	58	463	630	63	468	623	56	461
184	346	508	189	351	513	182	344	506	202	364	526	207	369	531	200	362	524	220	382	544	225	387	549	218	380	542
265	670	103	270	675	108	263	668	101	283	688	121	288	693	126	281	686	119	301	706	139	306	711	144	299	704	137
602	35	440	595	28	433	600	33	438	647	80	485	640	73	478	645	78	483	584	17	422	577	10	415	582	15	420
197	359	521	190	352	514	195	357	519	242	404	566	235	397	559	240	402	564	179	341	503	172	334	496	177	339	501
278	683	116	271	676	109	276	681	114	323	728	161	316	721	154	321	726	159	260	665	98	253	658	91	258	663	96
597	30	435	599	32	437	601	34	439	642	75	480	644	77	482	646	79	484	579	12	417	581	14	419	583	16	421
192	354	516	194	356	518	196	358	520	237	399	561	239	401	563	241	403	565	174	336	498	176	338	500	178	340	502
273	678	111	275	680	113	277	682	115	318	723	156	320	725	158	322	727	160	255	660	93	257	662	95	259	664	97
598	31	436	603	36	441	596	29	434	643	76	481	648	81	486	641	74	479	580	13	418	585	18	423	578	11	416
193	355	517	198	360	522	191	353	515	238	400	562	243	405	567	236	398	560	175	337	499	180	342	504	173	335	497
274	679	112	279	684	117	272	677	110	319	724	157	324	729	162	317	722	155	256	661	94	261	666	99	254	659	92

Solution by Daniele Degiorgi

Other than trivial magic subsquares of order 0 or 1, all embedded magic squares have orders which are a multiple of 3. We will say an embedded square of order n is of Type 1 if it consists of entries from n adjacent rows and n adjacent columns from our given 27×27 square. An embedded square of order n is of Type 2 if it consists of entries from n nonadjacent rows and n nonadjacent columns from the given square and the difference between the largest and smallest row index is the same as the difference between the largest and smallest column index. An embedded square of order n is of Type 3 if it consists of entries from n nonadjacent rows and n nonadjacent columns from the given square and it is not of Type 2.

We will call two embedded squares complementary if one consists of the entries in n designated rows and n designated columns from the original square and the other consists of entries from the remaining rows and columns. This means that the complementary embedded squares have orders n and $27 - n$.

Here is a count of all embedded magic squares which were found.

Single magic squares				
Order	Total	Type1	Type2	Type3
0	1	1	-	-
1	729	729	-	-
3	367	81	220	66
6	25092	-	4234	20858
9	827027	9	100562	726456
12	3526081	-	888867	2637214
15	2558237	-	814101	1744136
18	974440	-	401394	573046
21	23256	-	13030	10226
24	256	-	174	82
27	1	1	-	-
total 3-27	7934757	91	2222582	5712084
total	7935487	821	2222582	5712084

Complementary magic squares				
Order	Total	Type1	Type2	Type3
3	16	-	4	12
6	6332	-	1616	4716
9	188784	-	31132	157652
12	319953	-	104047	215906
15	319953	-	127037	192916
18	188784	-	79548	109236
21	6332	-	4126	2206
24	16	-	10	6
total	1030170	-	347520	682650

Order 3: One Type 1 magic square of order 3 consists of entries from the first three rows and first three columns of the given square. One example of Type 2 consists of the entries from rows 1, 4, 7 and columns 11, 14, 17. An example of Type 3 consists of entries from rows 3, 5, 7 and columns 1, 5, 9. Here are these examples.

$$
\begin{vmatrix} 638 & 71 & 476 \\ 233 & 395 & 557 \\ 314 & 719 & 152 \end{vmatrix}
\qquad
\begin{vmatrix} 8 & 1 & 6 \\ 3 & 5 & 7 \\ 4 & 9 & 2 \end{vmatrix}
\qquad
\begin{vmatrix} 314 & 712 & 150 \\ 228 & 392 & 556 \\ 634 & 72 & 470 \end{vmatrix}
$$

An example of a complementary square of order 3 consists of entries from rows and columns 9, 14, 19. Its complement, of order 24, consists of entries from rows and columns 1-8, 10-13, 20-27. Both are Type 2.

Order 6: An example of Type 2 consists of entries from rows and columns 2-4 and 6-8. An example of Type 3 consists of entries from rows 2-4, 6-8 and from columns 1,2,5,6,7,9. Here are these two squares.

395	557	226	550	231	393
719	152	307	145	312	717
66	471	635	473	637	70
714	147	311	149	313	718
67	472	639	477	632	65
391	553	234	558	227	389

233	395	388	550	231	555
314	719	712	145	312	150
633	66	68	473	637	475
309	714	716	149	313	151
634	67	72	477	632	470
229	391	396	558	227	551

One complementary square of order 6 consists of entries from rows and columns 8, 9, 13, 15, 19, 20. It and its complement are both of Type 2.

Order 9: One Type 1 example consists of entries from rows 1-9 and columns 1-9. One of Type 2 uses rows and columns numbered 1-3, 10-12 and 19-21.

638	71	476	575	8	413	620	53	458
233	395	557	170	332	494	215	377	539
314	719	152	251	656	89	296	701	134
593	26	431	611	44	449	629	62	467
188	350	512	206	368	530	224	386	548
269	674	107	287	692	125	305	710	143
602	35	440	647	80	485	584	17	422
197	359	521	242	404	566	179	341	503
278	683	116	323	728	161	260	665	98

Here is a Type 3 square using rows 7-9,13-15,19-21 and columns 6,8,9,12,14,16,19,20,22.

477	65	470	409	9	569	616	49	621
558	389	551	490	333	164	211	373	216
153	713	146	85	657	245	292	697	297
428	25	430	444	41	610	624	57	626
509	349	511	525	365	205	219	381	221
104	673	106	120	689	286	300	705	302
433	33	438	485	73	645	584	17	577
514	357	519	566	397	240	179	341	172
109	681	114	161	721	321	260	665	253

One Type 2 complementary square of order 9 consists of rows and columns 7-9,13-15,19-21.

Order 12: Here is a Type 2 square using entries from rows and columns 6-9,11,14-16,19-22.

149	313	718	151	651	653	86	250	291	696	129	293
477	632	65	470	4	9	414	569	616	49	454	621
558	227	389	551	328	333	495	164	211	373	535	216
153	308	713	146	652	657	90	245	292	697	130	297
505	186	348	510	368	361	523	204	224	386	548	217
509	187	349	511	363	365	527	205	219	381	543	221
104	268	673	106	687	689	122	286	300	705	138	302
432	587	20	425	40	45	450	605	625	58	463	630
433	600	33	438	80	73	478	645	584	17	422	577
514	195	357	519	404	397	559	240	179	341	503	172
109	276	681	114	728	721	154	321	260	665	98	253
437	601	34	439	75	77	482	646	579	12	417	581

Here is a Type 3 square using rows 6-9,11,14-16,19-22 and columns 2,5,8-9,13,15-16,18-22.

714	716	718	151	248	86	250	88	291	696	129	293
67	72	65	470	576	414	569	407	616	49	454	621
391	396	389	551	171	495	164	488	211	373	535	216
715	720	713	146	252	90	245	83	292	697	130	297
350	343	348	510	199	523	204	528	224	386	548	217
345	347	349	511	203	527	205	529	219	381	543	221
669	671	673	106	284	122	286	124	300	705	138	302
22	27	20	425	612	450	605	443	625	58	463	630
35	28	33	438	640	478	645	483	584	17	422	577
359	352	357	519	235	559	240	564	179	341	503	172
683	676	681	114	316	154	321	159	260	665	98	253
30	32	34	439	644	482	646	484	579	12	417	581

One complementary square of order 12 uses rows 6-9,11,14-16,19-22 and columns 5,7-19,12,13,15,16,19-21,23.

Order 15: This example of Type 2 consists of entries from rows 3, 5, 7–9, 12, 13, 15–17, 19–23 and from columns 3, 4, 7–9, 11, 14–16, 18–23 of the original square.

152	307	312	717	150	656	649	82	249	87	296	701	134	289	694
552	230	232	394	556	327	329	491	169	493	210	372	534	212	374
472	639	632	65	470	4	9	414	569	407	616	49	454	621	54
553	234	227	389	551	328	333	495	164	488	211	373	535	216	378
148	315	308	713	146	652	657	90	245	83	292	697	130	297	702
107	262	267	672	105	692	685	118	285	123	305	710	143	298	703
426	590	592	25	430	39	41	446	610	448	624	57	462	626	59
102	266	268	673	106	687	689	122	286	124	300	705	138	302	707
427	594	587	20	425	40	45	450	605	443	625	58	463	630	63
508	189	182	344	506	364	369	531	200	524	220	382	544	225	387
440	595	600	33	438	80	73	478	645	483	584	17	422	577	10
521	190	195	357	519	404	397	559	240	564	179	341	503	172	334
116	271	276	681	114	728	721	154	321	159	260	665	98	253	658
435	599	601	34	439	75	77	482	646	484	579	12	417	581	14
516	194	196	358	520	399	401	563	241	565	174	336	498	176	338

This Type 3 square uses rows 3, 6–9, 12–14, 16, 17, 19–23 and columns 3, 5, 6, 8–10, 12, 14, 16, 18–20, 22, 23, 25.

152	712	145	717	150	251	89	649	249	87	296	701	289	694	294
147	716	149	718	151	246	84	653	250	88	291	696	293	698	295
472	72	477	65	470	571	409	9	569	407	616	49	621	54	614
553	396	558	389	551	166	490	333	164	488	211	373	216	378	209
148	720	153	713	146	247	85	657	245	83	292	697	297	702	290
107	667	100	672	105	287	125	685	285	123	305	710	298	703	303
426	23	428	25	430	606	444	41	610	448	624	57	626	59	628
507	347	509	349	511	201	525	365	205	529	219	381	221	383	223
427	27	432	20	425	607	445	45	605	443	625	58	630	63	623
508	351	513	344	506	202	526	369	200	524	220	382	225	387	218
440	28	433	33	438	647	485	73	645	483	584	17	577	10	582
521	352	514	357	519	242	566	397	240	564	179	341	172	334	177
116	676	109	681	114	323	161	721	321	159	260	665	253	658	258
435	32	437	34	439	642	480	77	646	484	579	12	581	14	583
516	356	518	358	520	237	561	401	241	565	174	336	176	338	178

One complementary square of order 15 draws from rows 3, 6–9, 12–14, 16, 17, 19–23 and columns 3, 5–8, 10, 12, 14, 16, 18, 20–25.

Order 18: This Type 2 square uses rows and columns 4-12,16-24.

```
635   68  473  637   70  475  570    3  408  574    7  412  615   48  453  617   50  455
230  392  554  232  394  556  165  327  489  169  331  493  210  372  534  212  374  536
311  716  149  313  718  151  246  651   84  250  655   88  291  696  129  293  698  131
639   72  477  632   65  470  571    4  409  569    2  407  616   49  454  621   54  459
234  396  558  227  389  551  166  328  490  164  326  488  211  373  535  216  378  540
315  720  153  308  713  146  247  652   85  245  650   83  292  697  130  297  702  135
586   19  424  591   24  429  611   44  449  609   42  447  629   62  467  622   55  460
181  343  505  186  348  510  206  368  530  204  366  528  224  386  548  217  379  541
262  667  100  267  672  105  287  692  125  285  690  123  305  710  143  298  703  136
594   27  432  587   20  425  607   40  445  605   38  443  625   58  463  630   63  468
189  351  513  182  344  506  202  364  526  200  362  524  220  382  544  225  387  549
270  675  108  263  668  101  283  688  121  281  686  119  301  706  139  306  711  144
595   28  433  600   33  438  647   80  485  645   78  483  584   17  422  577   10  415
190  352  514  195  357  519  242  404  566  240  402  564  179  341  503  172  334  496
271  676  109  276  681  114  323  728  161  321  726  159  260  665   98  253  658   91
599   32  437  601   34  439  642   75  480  646   79  484  579   12  417  581   14  419
194  356  518  196  358  520  237  399  561  241  403  565  174  336  498  176  338  500
275  680  113  277  682  115  318  723  156  322  727  160  255  660   93  257  662   95
```

Here is a Type 3 square which uses rows 4-12,16-24 and columns 3,5-9,11-13,15-17,19-23,25.

```
471   68  473  637   70  475    3  408  572  410  574    7  615   48  453  617   50  619
552  392  554  232  394  556  327  489  167  491  169  331  210  372  534  212  374  214
147  716  149  313  718  151  651   84  248   86  250  655  291  696  129  293  698  295
472   72  477  632   65  470    4  409  576  414  569    2  616   49  454  621   54  614
553  396  558  227  389  551  328  490  171  495  164  326  211  373  535  216  378  209
148  720  153  308  713  146  652   85  252   90  245  650  292  697  130  297  702  290
431   19  424  591   24  429   44  449  604  442  609   42  629   62  467  622   55  627
512  343  505  186  348  510  368  530  199  523  204  366  224  386  548  217  379  222
107  667  100  267  672  105  692  125  280  118  285  690  305  710  143  298  703  303
427   27  432  587   20  425   40  445  612  450  605   38  625   58  463  630   63  623
508  351  513  182  344  506  364  526  207  531  200  362  220  382  544  225  387  218
103  675  108  263  668  101  688  121  288  126  281  686  301  706  139  306  711  299
440   28  433  600   33  438   80  485  640  478  645   78  584   17  422  577   10  582
521  352  514  195  357  519  404  566  235  559  240  402  179  341  503  172  334  177
116  676  109  276  681  114  728  161  316  154  321  726  260  665   98  253  658  258
435   32  437  601   34  439   75  480  644  482  646   79  579   12  417  581   14  583
516  356  518  196  358  520  399  561  239  563  241  403  174  336  498  176  338  178
111  680  113  277  682  115  723  156  320  158  322  727  255  660   93  257  662  259
```

The Type 2 example given above is complementary.

Order 21: This Type 2 example uses rows and columns 3-12,14,16-25.

```
152  307  712  145  312  717  150  251  656   89  649  249  654   87  296  701  134  289  694  127  294
471  635   68  473  637   70  475  570    3  408    5  674    7  412  615   48  453  617   50  455  619
552  230  392  554  232  394  556  165  327  489  329  169  331  493  210  372  534  212  374  536  214
147  311  716  149  313  718  151  246  651   84  653  250  655   88  291  696  129  293  698  131  295
472  639   72  477  632   65  470  571    4  409    9  569    2  407  616   49  454  621   54  459  614
553  234  396  558  227  389  551  166  328  490  333  164  326  488  211  373  535  216  378  540  209
148  315  720  153  308  713  146  247  652   85  657  245  650   83  292  697  130  297  702  135  290
431  586   19  424  591   24  429  611   44  449   37  609   42  447  629   62  467  622   55  460  627
512  181  343  505  186  348  510  206  368  530  361  204  366  528  224  386  548  217  379  541  222
107  262  667  100  267  672  105  287  692  125  685  285  690  123  305  710  143  298  703  136  303
507  185  347  509  187  349  511  201  363  525  365  205  367  529  219  381  543  221  383  545  223
427  594   27  432  587   20  425  607   40  445   45  605   38  443  625   58  463  630   63  468  218
508  189  351  513  182  344  506  202  364  526  369  200  362  524  220  382  544  225  387  549  218
103  270  675  108  263  668  101  283  688  121  693  281  686  119  301  706  139  306  711  144  299
440  595   28  433  600   33  438  647   80  485   73  645   78  483  584   17  422  577   10  415  582
521  190  352  514  195  357  519  242  404  566  397  240  402  564  179  341  503  172  334  496  177
116  271  676  109  276  681  114  323  728  161  721  321  726  159  260  665   98  253  658   91  258
435  599   32  437  601   34  439  642   75  480   77  646   79  484  579   12  417  581   14  419  583
516  194  356  518  196  358  520  237  399  561  401  241  403  565  174  336  498  176  338  500  178
111  275  680  113  277  682  115  318  723  156  725  322  727  160  255  660   93  257  662   95  259
436  603   36  441  596   29  434  643   76  481   81  641   74  479  580   13  418  585   18  423  578
```

This Type 3 example uses rows 3-12,14,16-25 and columns 2,3,5-10,12-16,18-23,25,26.

719	152	712	145	312	717	150	251	89	244	649	82	249	87	296	701	134	289	694	294	699
66	471	68	473	637	70	475	570	408	572	5	410	574	412	615	48	453	617	50	619	52
390	552	392	554	232	394	556	165	489	167	329	491	169	493	210	372	534	212	374	214	376
714	147	716	149	313	718	151	246	84	248	653	86	250	88	291	696	129	293	698	295	700
67	472	72	477	632	65	470	571	409	576	9	414	569	407	616	49	454	621	54	614	47
391	553	396	558	227	389	551	166	490	171	333	495	164	488	211	373	535	216	378	209	371
715	148	720	153	308	713	146	247	85	252	657	90	245	83	292	697	130	297	702	290	695
26	431	19	424	591	24	429	611	449	604	37	442	609	447	629	62	467	622	55	627	60
350	512	343	505	186	348	510	206	530	199	361	523	204	528	224	386	548	217	379	222	384
674	107	667	100	267	672	105	287	125	280	685	118	285	123	305	710	143	298	703	303	708
345	507	347	509	187	349	511	201	525	203	365	527	205	529	219	381	543	221	383	223	385
22	427	27	432	587	20	425	607	445	612	45	450	605	443	625	58	463	630	63	623	56
346	508	351	513	182	344	506	202	526	207	369	531	200	524	220	382	544	225	387	218	380
670	103	675	108	263	668	101	283	121	288	693	126	281	119	301	706	139	306	711	299	704
35	440	28	433	600	33	438	647	485	640	73	478	645	483	584	17	422	577	10	582	15
359	521	352	514	195	357	519	242	566	235	397	559	240	564	179	341	503	172	334	177	339
683	116	676	109	276	681	114	323	161	316	721	154	321	159	260	665	98	253	658	258	663
30	435	32	437	601	34	439	642	480	644	77	482	646	484	579	12	417	581	14	583	16
354	516	356	518	196	358	520	237	561	239	401	563	241	565	174	336	498	176	338	178	340
678	111	680	113	277	682	115	318	156	320	725	158	322	160	255	660	93	257	662	259	664
31	436	36	441	596	29	434	643	481	648	81	486	641	479	580	13	418	585	18	578	11

The Type 2 example given above is complementary.

Order 24: A Type 2 complementary square of order 27 uses rows and columns 2-13 and 15-26. A Type 3 square of this order uses rows 2-13, 15-26 and columns all columns other than 3, 13 and 26.

Editor's Commentary

Clarence Gipbsin sent me this magic square and I posed a problem based on it. My statement of the problem was purposely vague because I wanted to see how readers interpreted my question and what discoveries they would make. While working to understand Gipbsin's construction method, I became aware of the embedded magic squares which Degiorgi calls Type 1 and of those of Type 2 which draw from equally space rows and columns of the given square. An understanding of these subsquares is the key to understanding Gibpsin's construction method, after which it is easy to see that his method can easily produce magic squares of unlimited size. I was pleasantly surprised by Degiorgi's discovery of the embedded squares which he calls Type 3 and by his complementary subsquares. I do know of one other way in which magic squares are embedded in the given square. Rather than revealing it at this time, I will promise to publish any further discoveries from readers whose interest might be aroused by what has been found to this point. I believe that Degiorgi's three types of squares are related to his experience in writing a program to search for them. Perhaps some other classification would give even more insights.

2. Pentomino Words II by Brian Barwell, Hampton, Middlesex, UK

Figure 2 shows the twelve pentominoes and their associated reference letters. A pentomino word is one containing only the letters *F, I, L, N, P, T, U, V, W, X, Y* and *Z* such that the pentominoes corresponding to the letters in the word can be arranged to form a rectangle. Repeated letters are allowed. Figure 3 shows an example of such a rectangle made from the letters of the six-letter word PULPIT.

Find a pentomino word containing more than six letters. What is the longest pentomino word?

Figure 2

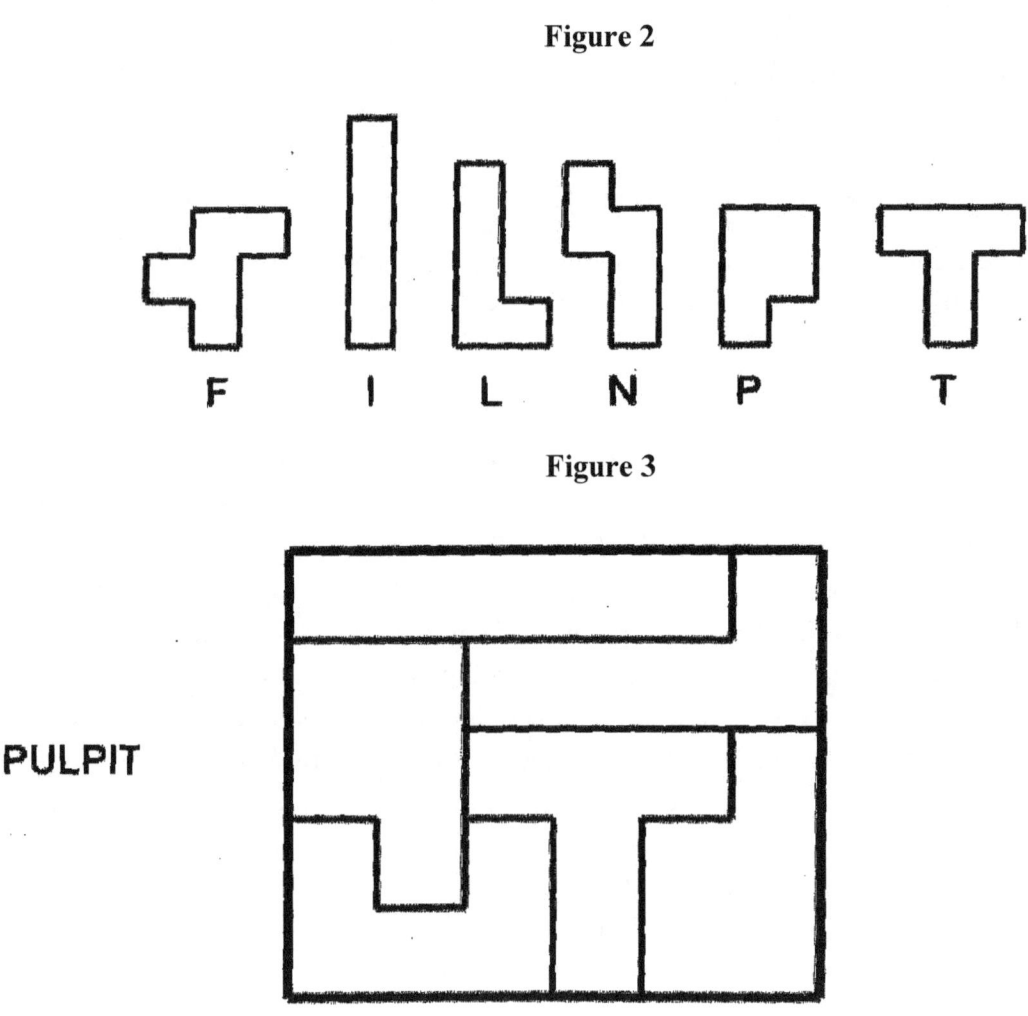

Figure 3

PULPIT

Solution by Andy Pepperdine

The 11-letter word UNPITIFULLY can be packed into an 11×15 rectangle as shown in Figure 4.

Solution by Daniele Degiorgi

All of the following are pentomino words: FINNILY, NIPPILY, NULLIFY, PITIFUL, PUFFILY, PULPILY, PULVINI, TINNILY, UNFITLY, WILLFUL, WITTILY, FITFULLY, FLINTILY, FLUFFILY, FUTILITY, PULVILLI, TITTUPPY, WILFULLY, PITIFULLY, UNFULFILL, WILLFULLY .

Figure 4

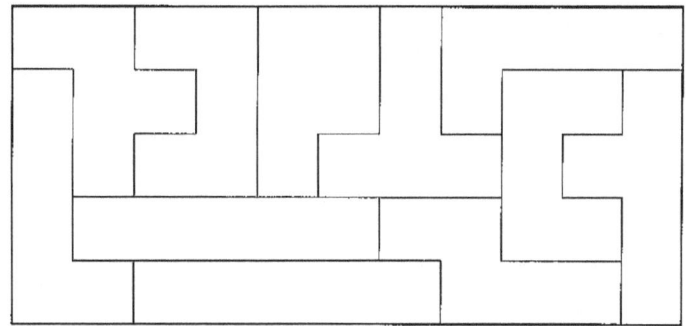

3. Comeback 2511 by Paul E. Boymel, Potomac, MD

Problem 2511 (*JRM 30*:2, p. 148, *JRM 31*:2, p. 145) asked for an arrangement of the integers 1 through 25 in a square array so that each number except 1 and 2 is the sum of two of its neighbors, orthogonal or diagonal. This problem has been found to have roughly 5000 solutions.
a. Find a solution where the number 1 is on the outside edge of the square.
b. Find the solution having the largest possible sum of any row, column or diagonal.
c. Find the solution having the smallest possible sum of any row, column or diagonal.

Solution by Daniele Degiorgi

a. There is one such solution.

21	11	1	13	18
10	4	3	5	12
20	6	2	7	19
14	8	15	9	16
22	23	17	24	25

b. There are twelve solutions having the maximum sum 115. It never appears as a diagonal sum. Here is one.

25	15	11	10	16
24	1	4	6	18
23	9	3	2	20
22	8	5	7	14
21	13	17	12	19

c. There are 73 solutions having the minimum possible sum which is 26. Here is one.

23	11	10	19	24
12	1	4	9	15
13	16	3	6	17
18	5	2	8	14
25	20	7	21	22

Making Connections—the Antidote to Chaos

Kate Jones

Previously I wrote about games and puzzles with "row by row" arrangements and goals. Now let's look at some games that connect the ends of rows to a start and end position, even like a complex spider's web anchored in several places, or even to each other as loops. Bridges, roads and rails, phone wires, cables, a dog on a leash, a line of code, a sentence spoken by one and heard by another, a simple handshake… all these are connections that hold life and systems together. How natural that we'd build the concept into games. After all, games are microcosms of world and mind.

A great resource of historical information about connection games is the Boardgamegeek website: http://www.boardgamegeek.com/boardgamefamily/7345/connection-games. It lists hundreds of connection games, including Hasbro's Bridg-It (1960), Twixt (1962), and Piet Hein's Hex (1942), most of them now out of print. The list shows only a couple of Kadon's many connection games, mainly from fans reviewing them, since Kadon products are not sold in stores. To make up for this shortage, we'll take a look at some others here. You can see all of them and more on the Gamepuzzles website, www.gamepuzzles.com.

Paths and Loops

With games that consist of lines imprinted on their pieces, forming connections is a natural, not only intricate but also beautiful. All those shown here can have many different arrangements.

The **Kaliko** set was created by Charles Titus and Craige Schensted (who later changed his name to Ea) and produced under the name *Psyche-Paths* in the 1970s, consisting of 85 hexagonal tiles with every combination of paths in 3 colors. The goal is to form connections of paths in matching colors and to close loops for extra points. Each turn begins by connecting two loose ends to each other. The original sets were made of clear acrylic. Kadon obtained the rights to Kaliko in 1991 and introduced the wood version in 2001.

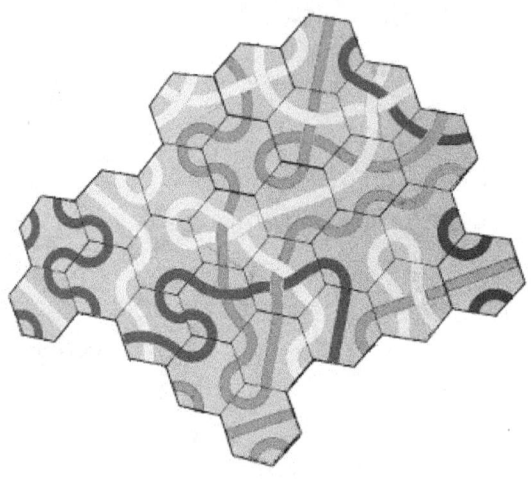

Arc Angles consist of 25 curved wedges, related directly to the golden ratio, being one-fifth of a circle. The arc-shaped paths are all the different combinations of connecting five points along each edge. The tiles adjoin with in and out bends and can form five separate circles or even one closed figure. Created by Kate Jones in 2005.

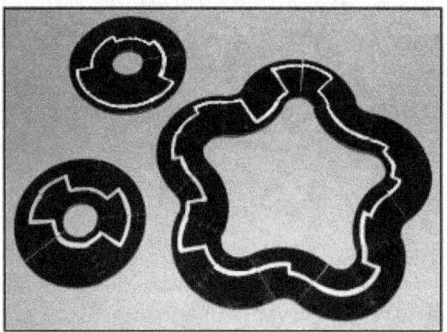

Fractured Fives has 5 reversible square tiles with images that match on their edges. One side is a system of tangled ropes, and the goal is to connect a single rope that traverses all 5 squares in any of the 12 different arrangements they can form ("pentominoes"). The image at left is *not* such a solution. Invented in 1986 by Kate Jones, published in 2011.

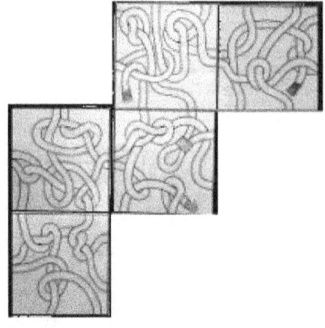

Dezign-8 contains 64 square tiles inlaid with path segments with from 1 to 4 exits. Connect them into a single network that lets you travel from any tile to any other (one solution is shown). Interestingly, up to 19 separate path groupings can be formed, always with an equal number of loops built in. Symmetry patterns make this into a dramatic work of art as well as an endlessly interesting strategy game of connectivity. Launched in 2000.

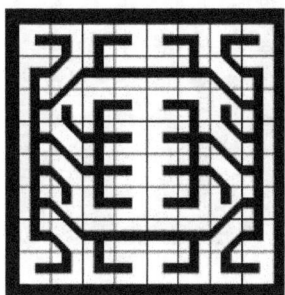

Hexmozaix and ***Hexmozaix II*** have hexagonal tiles inlaid with every combination of 3 colors forming diamonds and chevrons. Connect them to form matching-color paths that score points for their length. How long can you go? Invented by Charles Butler, developed by Kate Jones, and published by Kadon since 1988-1989.

ChooChooLoops is a gamepuzzle of "track" pieces formed of combinations of quarter arcs, joined in groups of 1 through 4 in all their different shapes. The track segments are to be connected into paths and loops that thread their way around the 36 permanent black "islands". What is the maximum number of islands you can enclose in a single loop? All but the ring piece can form one continuous convoluted path. A special challenge is to use just one each of the ten distinct shapes and build a connection from top to bottom of a 3x6 area of the tray. Designed by Kate Jones.

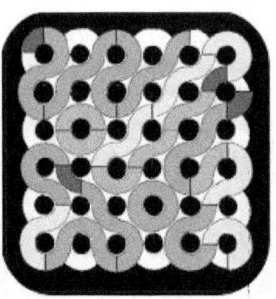

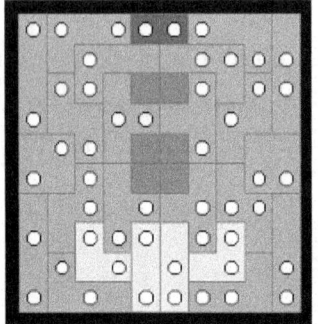
Fill-Agree (left), by Kate Jones (1996), and *L-Sixteen* (right), by Nevzat Moraç and Kate Jones (2013), are a change of pace. Each piece is unique and sports some holes. Connect the *holes* into a single continuous maze (one solution is shown at right). Holes on left are not connected. Can you solve it?
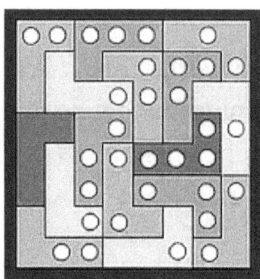

Flagstones and Bridges

Tiling sets with all different members of a certain concept, by shape or color, will readily form many designs by matching adjacent pieces. We can require them in addition to form long paths that link distant points or form complete territories. Here are some examples from Kadon's line:

Quintapaths consists of 20 "sticks", each a distinct arrangement of 0 to 5 black squares on a 1x5 white rectangle. Invented by world-renowned puzzle master Scott Kim in 1969, when he was only 14 years old! Further developed by Kate Jones and made by Kadon since 1999. Connect the black sections to form complete paths and loops and symmetries.

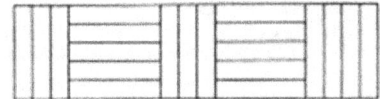

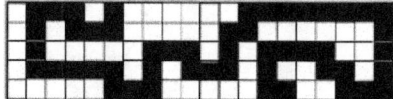 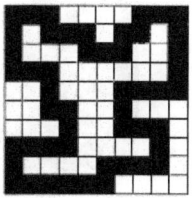

Intarsia has just two kinds of reversible tiles, 16 of each, and fill the hexagonal tray in septillions of ways, including breathtakingly beautiful symmetry patterns. A few rare solutions form a single path of one color (shown). Among several games is "Chains", where two players, playing black and white, seek to connect non-adjacent sides of the tray with paths of their own color. I nvented by Henrik Morast, developed by Kate Jones and published by Kadon since 2008.

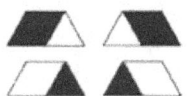 

Bear Hugs Jr. is a charming teaching tool for ages 3 and up, featuring 16 teddybears in all-different poses that mimic the 16 permutations of DaVinci's famous man. One of the bears' connection goals is to form them into a circle where neighboring bears match by arm and leg, and the circle forms a closed loop (shown at right).

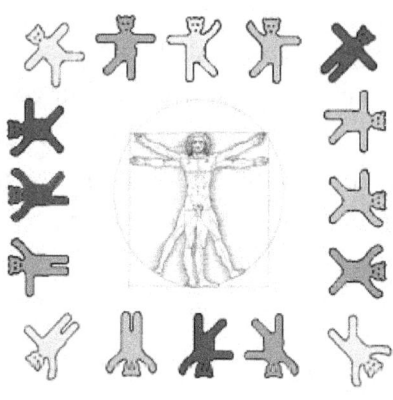

Kadon also makes a 36-bear set, *Bear Hugs*, that likewise can form a full-loop connection. Both sets come in trays, 4x4 and 6x6 with 4 and 6 colors, respectively, that represent Latin Squares. Created by Kate Jones.

Trifolia has 24 unique triangles with four kinds of edge. While the hexagon has 2.3 million solutions, only 173 have both colors fully connected. Even rarer: congruent color regions (shown at right). Created by Kate Jones.

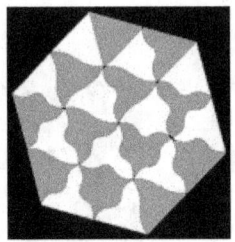 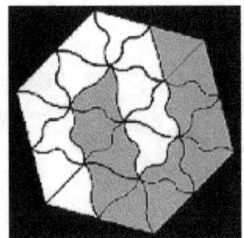

Hopscotch is a polyform set of squares of orders 1 through 4 ("polyhops") that plot on a grid where each row is displaced by half a space. In "Gardenpath", two players strive to build connections for maximum length and largest enclosed area. Take care that your path connects back into a loop. Invented by Thomas Atkinson, developed by Kate Jones. Published in 2011.

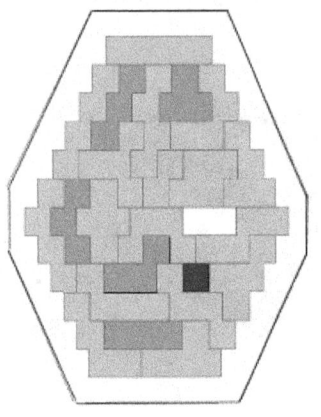 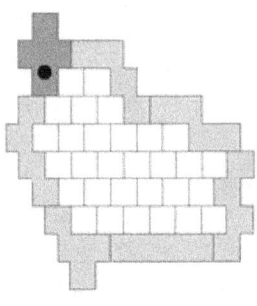

Rombix is an unusual type of connection: "ladders" that join opposite edges of the tray. There are four pieces each of the four colors, and any color can form the top-to-bottom connecting ladder. In fact, there are 96 different ways to fit a ladder across the center line. Invented by mathematician Alan Schoen, with additional development by Kate Jones. This venerable set has been in print since 1992, with a fabulous pedigree: A "Games 100" selection for 1993, and Alan was an associate of Bill Perk, who was an associate of Buckminster Fuller in Carbondale, IL.

Other Kadon sets with connection themes are *Hexnut Jr.*, *Sextillions*, *Quintachex*, *Vee-21*.

Trails of Stones

Classic strategy games, like *Go*, have players placing stones upon a grid until one or the other achieves a winning position, generally to connect opposite edges. In these two games, created by Titus and Schensted, the goal is to connect at least 3 edges.

The Game of Y has a curvy triangular board, somewhat like a flattened geodesic dome, and players connect all 3 sides to win. Stones go on intersections. A corner counts for both sides. Always a winner, never a draw. Published in book form by Neo Press in 1975, the physical version has been made by Kadon since 1993.

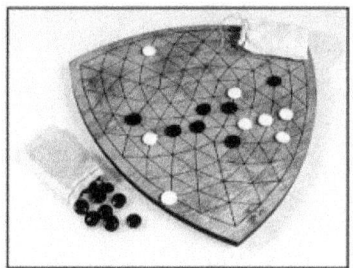

*Star was introduced in 2004, a more complex theme of connecting edges. Stones occupy spaces. Players seek to "own" edge spaces by occupying or surrounding them. The concentric pentagon board allows for 3 sizes of play. Unique scoring system favors connecting one's stones into as few separate groups as possible, so defense is crucial. The central star is a bridge for both players.

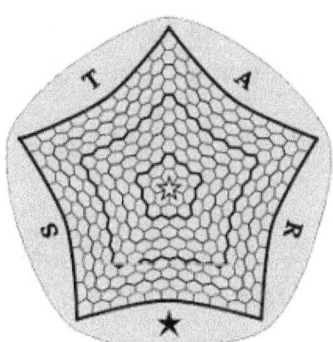

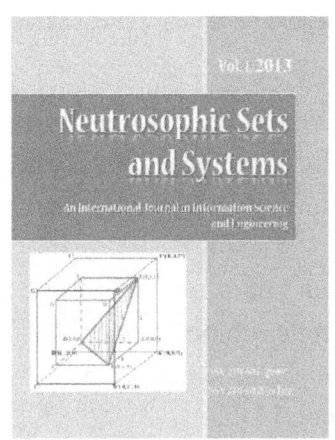

Editor-in-Chief:

Prof. Florentin Smarandache

Department of Mathematics and Science

University of New Mexico

705 Gurley Avenue

Gallup, NM 87301, USA

E-mail: smarand@unm.edu

Home page:
http://fs.gallup.unm.edu/NSS

Associate Editors:

Dmitri Rabounski and Larissa Borissova, independent researchers.

Said Broumi, Univ. of Hassan II Mohammedia, Casablanca, Morocco.

A. A. Salama, Faculty of Science, Port Said University, Egypt.

Yanhui Guo, School of Science, St. Thomas University, Miami, USA.

Francisco Gallego Lupiañez, Universidad Complutense, Madrid, Spain.

Peide Liu, Shandong Universituy of Finance and Economics, China.

Pabitra Kumar Maji, Math Department, K. N. University, WB, India.

S. A. Albolwi, King Abdulaziz Univ., Jeddah, Saudi Arabia.

Mohamed Eisa, Dept. of Computer Science, Port Said Univ., Egypt.

Neutrosophic Sets and Systems has been created for publications on advanced studies in neutrosophy, neutrosophic set, neutrosophic logic, neutrosophic probability, neutrosophic statistics that started in 1995 and their applications in any field, such as the neutrosophic structures developed in algebra, geometry, topology, etc.

The submitted papers should be professional, in good English, containing a brief review of a problem and obtained results. Neutrosophy is a new branch of philosophy that studies the origin, nature, and scope of neutralities, as well as their interactions with different ideational spectra.

This theory considers every notion or idea <A> together with its opposite or negation <antiA> and with their spectrum of neutralities <neutA> in between them (i.e. notions or ideas supporting neither <A> nor <antiA>). The <neutA> and <antiA> ideas together are referred to as <nonA>.

Neutrosophic Set and Logic are generalizations of the fuzzy set and respectively fuzzy logic (especially of intuitionistic fuzzy set and respectively intuitionistic fuzzy logic). In neutrosophic logic a proposition has a degree of truth (T), a degree of indeterminacy (I), and a degree of falsity (F), where T, I, F are standard or non-standard subsets of $]^{-}0, 1^{+}[$.

Neutrosophic Probability is a generalization of the classical probability and imprecise probability.

Neutrosophic Statistics is a generalization of the classical statistics.

What distinguishes the neutrosophics from other fields is the <neutA>, which means neither <A> nor <antiA>. <neutA>, which of course depends on <A>, can be indeterminacy, neutrality, tie game, unknown, contradiction, ignorance, imprecision, etc.

All submissions should be designed in MS Word format using our template file:

http://fs.gallup.unm.edu/NSS/NSS-paper-template.doc

A variety of scientific books in many languages can be downloaded freely from the Digital Library of Science:

http://fs.gallup.unm.edu/eBooks-otherformats.htm

To submit a paper, mail the file to the Editor-in-Chief. To order printed issues, contact the Editor-in-Chief. This journal is non-commercial, academic edition. It is printed from private donations.

Information about the neutrosophics you get from the UNM website:

http://fs.gallup.unm.edu/neutrosophy.htm

The home page of the journal is accessed on

http://fs.gallup.unm.edu/NSS

BOOKS IN RECREATIONAL MATHEMATICS BY CHARLES ASHBACHER AND ASSOCIATES

Topics in Recreational Mathematics 1/2015 ISBN 978-1507603215

Topics in Recreational Mathematics 2/2015 ISBN 978-1508617099

Topics in Recreational Mathematics 3/2015 ISBN 978-1511641005

Topics in Recreational Mathematics 4/2015 ISBN 978-1514317518

Topics in Recreational Mathemastics 5/2015 ISBN 978-1519115676

Topics in Recreational Mathematics 1/2016 ISBN 978-1530003655

Alphametics as Expressed in Recreational Mathematics Magazine ISBN 978-1508538134

Ten Year Cumulative Index to the Journal of Recreational Mathematics, edited by Joseph S. Madachy and Charles Ashbacher ISBN 978-1508936800

Alphametics Expressing Thoughts From the Star Trek Original Series ISBN 978-1512152784

Mathematical Cartoons ISBN 978-1514207130

Solved Problems in Statistical Inference ISBN 978-1515215622

Associates

Artist Catie Ribble

Editor Rachel Pollari

Editor Jennifer Corrigan

Artist Jenna Richardson